$9.25

# ROOFERS HANDBOOK

## By William E. Johnson

**Craftsman Book Company**
542 Stevens Avenue
Solana Beach, California 92075

**Library of Congress Cataloging in Publication Data**

Johnson, William Edgar, 1938-
  Roofers handbook.

  Includes index.
  1. Roofing, Bituminous--Handbooks, manuals, etc.
  2. Shingles--Handbooks, manuals, etc. I. Title.
TH2441.J63      695'.1      76-5875
ISBN 0-910460-17-5

© 1976 Craftsman Book Company

Illustrations by Laura Knight and the author
Cover design by Bill Grote Advertising/**Graphics**

Sixth Printing 1986

# Contents

Your Future In a Changing Roofing Business ---5

Chapter 1, Tools and Equipment------------------8
   Ladder loaders ----------------------------8
   Small tools -------------------------------9
   Roofer's hatchet ------------------------10
   Roof jacks-------------------------------11

Chapter 2, Basics of Shingle Roofing ----------13
   Shingle types ---------------------------13
   Beginning a roof -----------------------13
   The importance of following directions -----15
   Roofing terms --------------------------16

Chapter 3, Three Tab Shingles -----------------18
   Six inch pattern, stair step, new roof --------18
   Five inch pattern, stair step, new roof ------21
   Six inch pattern, straight-up, new roof -----23
   Four inch pattern -----------------------24
   Straightening up horizontal lines -----------26
   Eliminating the starter course overlap -----27
   Offsetting the overlap: Butting up ----------27
   Starting a butt-up roof -------------------29
   Starting in the middle: Reroofing
     straight-up---------------------------31
   Gravel roof tear-off and reroof ------------33
   Gravel tear-off -------------------------35

Chapter 4, T-Lock Roofing Shingles -----------36
   Tips on T-locks -------------------------38
   Straightening T-locks -------------------41

Chapter 5, Shingle Tie-Ins --------------------43
   Tie-ins with uneven eaves ----------------45
   Butt-up tie-ins using six inch pattern -------47
   T-lock tie-ins on new and old roofs---------47
   Shingling around a dormer, six
     inch pattern--------------------------53
   Shingling around a chimney---------------54
   Difficult T-lock tie-ins -------------------54
   Five inch pattern tie-ins using three tabs --56

Chapter 6, Flashing Fireplaces and Chimneys 60
   Flashing on reroofs ---------------------62

Chapter 7, Valleys ---------------------------65
   The full lace valley----------------------65

   The half lace valley----------------------66
   W shaped formed metal valley -------------67
   The smooth valley ----------------------68
   Valley tips ------------------------------69
   Waterproofing a blind valley----------------71
   Improper valley material------------------72
   Replacing with W shaped formed valley ---74
   Replacing with 90 pound roll roofing-------74
   Cutting in a valley on an addition -----------76

Chapter 8, Ridges --------------------------77
   Cutting hip ridges out of three tabs--------78
   Tips on ridge work----------------------79

Chapter 9, Tips For Professional Roofers-------83
   Shingle selection-----------------------84
   Felt underlay --------------------------86
   Flashing around vents -------------------86
   Step flashing --------------------------90
   Spot repairs ---------------------------94
   Replacing individual shingles---------------94
   Metal edging --------------------------96
   Trimming rakes ------------------------98
   Tear off tips ---------------------------99
   Roll roofing tips -----------------------100

Chapter 10, Handling Leaks -------------------102
   Valley leaks ---------------------------103
   Vent and flashing leaks ------------------106
   Other leaks ---------------------------110

Chapter 11, Wood Shingles and Shakes -------112
   Tools and equipment -------------------112
   Material selection ---------------------113
   Ventilation ----------------------------114
   Estimating ----------------------------114
   Application of wood shingles -------------115
   Valleys --------------------------------118
   Shingling on steep roofs ------------------119
   Flashing wood shingles -----------------120
   Ridge ---------------------------------122
   Dutch weave pattern -------------------124
   Applying shakes -----------------------125
   Valleys --------------------------------128
   Vents ---------------------------------130
   Hips and ridges -----------------------131

**Chapter 12, Operating a Roofing Company** --134
    Roof covers ---------------------------------------135
    Estimating roof areas ------------------------136
    Labor costs --------------------------------------138
    Direct overhead --------------------------------139
    Indirect overhead ------------------------------140
    Profit ---------------------------------------------140
    Crew and equipment management --------141
    Training programs ----------------------------141
    Starting a new company ---------------------142

**Chapter 13, Selling Your Services** --------------144
    Your advertising budget ---------------------144
    Planning your advertising -------------------145
    Effective advertising --------------------------145
    Newspapers ---------------------------------------145
    Billboards, posters, signs --------------------146
    Matchbooks, pencils, rulers -----------------147
    Letters --------------------------------------------147
    Printed matter ----------------------------------149
    Following up on prospects ------------------149
    Selling the prospect --------------------------149
    Salesmanship ------------------------------------150
    Approaching the prospect -------------------151
    Demonstrating your service -----------------151
    Answering questions and objections ------153
    Closing --------------------------------------------154
    When to excuse yourself---------------------154
    Sample contracting forms --------------------155

**Reference Section** ----------------------------------159
    Bird roofing materials--------------------------160
    Celotex roofing materials---------------------162
    Certain-teed roofing materials --------------167
    Flintkote roofing materials-------------------172
    Fry roofing materials --------------------------174
    GAF roofing materials -------------------------177
    Johns-Manville roofing materials ----------181
    Owens Corning roofing materials ----------184
    Tamko roofing materials---------------------185

**Index** -----------------------------------------------------188

# Your Future in a Changing Roofing Business

In the distant past a good roof meant just having some shelter overhead. Our ancestors used sod, animal skins, straw, reeds or pieces of flat boards. No doubt, the people who lived under these shelters didn't expect as much from a roof as we do today. A modern shingle roof can be expected to last 15 to 20 years and should keep the interior dry and comfortable for at least that long.

While shingles in one form or another are quite old, asphalt shingles were not widely available until around 1870. But asphalt itself was used as a building material as early as 3,000 B.C. While asphalt has always been one of the least expensive and most durable roofing materials, only in the last ten years have asphalt shingles become an important design feature of the structure. Manufacturers now offer a wide variety of earth tones and have many irregular textures that look better and wear longer similar to wood shingle or shake roofs. While most of the credit for these developments must go to the shingle manufacturers, some of the credit must go to the public in general. Modern homeowners want something better looking and builders and architects are putting more emphasis on roof styles. The trend in modern home design is toward dramatic roof lines that emphasize the roof covering and away from ranch style built-up roofs. Ten years ago homeowners were not concerned about the roof as long as the ceiling didn't develop water spots. Most homeowners then probably could not tell you the color of the roof cover on their home unless they went outside to check it. Today this is changing. Look at the new homes that are being built today and you can see the changes. Take a good look at these new homes from a distance. About half of what you see will be the roof. Since so much of the exterior is the roof, why shouldn't it look nice? Notice the style, the color and the pattern. As a roofer you will also notice the edges, the valleys and the ridges, and how straight the shingle pattern is.

The change in our attitude about roofing is opening up many opportunities for professional roofers. Chapters 12 and 13 explain how professional roofers make a living. It is enough to point out here that the demand for roofing services is sure to increase during the next decade. During the late 1950's and 1960's millions of new housing units were completed.

Nearly all of these homes will have to be reroofed during the next decade. Today, nearly 80% of all the work small roofing contractors do is reroofing. This figure is bound to increase during the 1980's. Unlike buying a new car or a larger home or remodeling a kitchen, recovering a badly leaking roof can not be postponed indefinitely. But most important from a sales standpoint, putting on a new roof is now far more than just restoring the roof surface to a "like new" condition. The new composition shingles look better, wear longer, and are more resistant to wind and hail damage than the old asphalt shingles. A new roof is now a definite home improvement that adds to the resale value like a room addition or remodeled bathroom. Understanding this, it's easy to see a bright future for professionals in the roofing industry.

Even though there are many opportunities opening for roofers, you should not assume that almost anyone who calls himself a roofer can make a good living. Great technical skill, knowledge and craftsmanship are needed to put on a durable, attractive, dependable roof. This book is written to help anyone develop the competence a roofer needs to do first class, professional work. The information you find in the following chapters is the product of many years of roofing experience. But the roofing trade is broad and varied and no one has all the answers to all questions. However, the procedures illustrated here are good roofing practice and you can rest assured that if you follow this book you will put on a good roof.

It may seem obvious, but every roofer should understand the importance of every shingle he places. The roof will be subject to intense heat, driving rain, wind and even water under pressure. In cold climates snow will build up on the roof. Though the outside temperature may not be high enough to melt the snow, there will be enough heat escaping through the attic to melt the snow touching the shingles. This causes water to run down to the eave where it meets cold air and freezes, forming an ice dam. The water that is still running down the roof starts backing up under pressure until eventually it may find a way under the shingles and into the house. This is what we call a "freezeback" and it can cause a lot of damage to the interior and exterior of a house.

Some homeowners reroof their own homes and claim a sound roof even though they don't know anything about roofing. Fortunately for these homeowners, you really have to do something bad to make a reroof leak enough so that it can be seen from the inside. Some of the more serious roof leaks don't show any evidence of leaking until it is too late. A leak doesn't have to be obvious to be a bad leak. About 75 percent of all houses have a leak of some type somewhere or will begin leaking long before the anticipated useful life of the roof has passed. The small leak may allow a little water to trickle down inside a wall. The subfloor absorbs all the water and no moisture shows. But the wet wood will rot eventually and result in an expensive repair job. Some roofs have a small leak around a vent that only soaks a couple of roof deck boards. These boards rot and lose their ability to absorb water. Then the rafter will start to soak up excess water and rot. A roof is like a paint job; if moisture can get under the surface it will deteriorate rapidly. A leaking roof will probably be slower to fail completely than a paint job but it usually is much more costly to repair. The last three jobs this author completed before compiling this book ended up being costly repair jobs. One of these roofs was a tear off job on a well maintained ranch style house in a suburban neighborhood. We ended up replacing over 100 dollars worth of lumber. Naturally the labor cost was much more. Fascia, trim, sheathing boards, and four 2x6 rafters were replaced. The homeowner was surprised because there was no indication of a leak. After the roof was removed you could hardly walk around on the roof without stepping in a hole. In this case, as in most similar situations, it was the fault of the roofer.

We are not saying that 75 percent of all roofers are careless or poor roofers. There are thousands of highly experienced and quite capable roofing specialists. However, most of these roofers do not do all of their own work. They hire a helper or a trainee. While the roofer is solely responsible for everything that happens on the roof, he cannot watch every nail and shingle that is used. On an average 3-tab roofing job on a 2,000 square foot roof there are about 5,000 pounds of shingles (about 1,600 shingles) and about 6,500 nails. Also, there will be several vents to handle correctly, a couple of valleys to lay and a chimney to flash around. If it is a new roof there will be a 15 pound layer of roofing felt to nail down. Of course, the roofer will want to finish the house in one day if he has a helper and wants to make money. He sometimes will try to apply the shingles when the wind is blowing too hard or the temperature is too high or too low, or in some condition or manner against manufacturer's recommendations. Unless great care is used or the roofer is very lucky, the result is going to be at least one small leak.

Roof leaks are not always the fault of the roofer. Sometimes it may be the fault of the builder or carpenter. A leak can be the result of a crooked valley that a carpenter thinks the roofer can conceal. Sometimes a certain area may be trimmed out with wood trim or siding before the roofer has a chance to flash it with metal. Sometimes an architect designs a tri-level with an upper deck that emptys large amounts of rain water at an angle onto a lower deck. The result will almost surely be a problem roof no matter what the roofer does. Quite often the builder himself is rushed and insists that the roof be applied before the plumber or heating man has had a chance to get vents up through the roof deck. Sometimes a mason finishes with the chimney after the roof is installed and loose sand and mortar are washed under the shingle tabs. Eventually the roof leaks. A homeowner may install a television antenna in a poor location or install christmas decorations. Neighborhood kids retrieving toys will cause leaks, particularly on dry wood shingles or cold, brittle asphalt shingles. This could go on and on but you understand by now that installing a durable, sound roof is not an easy task, even under ideal conditions.

In summary, as a roofer you follow a demanding and rewarding trade in a changing industry. You will achieve personal satisfaction in proportion to your ability to absorb technical knowledge and professional skill as a roofer.

# Chapter 1
# Tools and Equipment

Every craftsman needs the right equipment, in good condition, to do first class work. This is especially true of the tools of the roofer's trade. Specialized tools are expensive but many will more than repay you the initial cost.

### Ladder Loaders

Figure 1-1 shows a ladder loader with a 200 pound capacity. This limit should not be exceeded for any reason. Proper care is very important for this loader as with any equipment. Most ladder loaders are aluminum and can be damaged easily. The pulleys and bearings must be kept clean and well greased. Figure 1-2 shows the platform assembly that carries the shingles up the ladder. The small home-made extension with two roller wheels on the front side is a big help for unloading T-lock shingles. Without this extension, the tabs of the T-locks will hang down just enough to prevent unloading. Proper positioning of the ladder is extremely important for smooth unloading. When the ladder is the right height the six foot angle guide will slope slightly down to the roof. To make this height adjustment easier, the ladder sections for this loader were cut so that they could be set up in two foot increments. Sections 16, 10, 8, 4 and 2 feet long will give you the right combination for about any roof. Even then you may need to set the loader on a couple of bundles of shingles, as in Figure 1-1, if the ladder is a little short. If the ladder is too long the angle guide will be at too steep an angle and the shingles will quickly slide to the roof, putting all of the weight on the rafters. This can

Aluminum ladder loader
Figure 1-1

*Tools and Equipment*

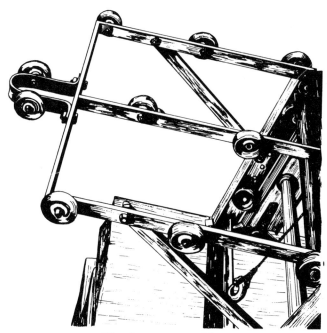

Platform assembly
Figure 1-2

Loaded platform assembly
Figure 1-3

be corrected by placing one or two bundles of shingles under the angle guide at the discharge end.

In Figure 1-3 you can see two bundles on the platform assembly ready to be hoisted up the ladder. Note that the top bundle is back from the ladder about five inches. This will take some of the weight off the bottom bundle in unloading when it hits the rollers on the angle guide and smooth the unloading of shingles. Take a couple of extra minutes to set this ladder loader up correctly. A good set-up will save much more time than it takes.

If you carry this loader on racks on top of a truck you should protect the loader by wrapping the load bearing surfaces of the rack with ¼ inch nylon cord. This will keep the ladders from wearing out and eliminate any annoying rattle. Many roofers carry the six foot angle guide on top of the ladders. The guide can bounce around and wear holes in the ladder rungs. Avoid this by tying a one inch strip of old auto tire rubber on the ladder rungs in four places, two at each end of the angle guide. Heavy rubber tie-downs will help keep the ladders from bouncing on top of the racks. Your ladder loader will give good service for many jobs if you take good care of it.

Small Tools

The tools that you need to apply shingles depend on your preferences and the type of jobs you are working on. When laying three tab shingles or strip shingles, many roofers use a nail "stripper" as shown in Figure 1-4. This is a specially made stainless steel box that is open at the top and has either two or three channels in the bottom. These channels are just wide enough for the nails to fall through and hang by the heads. At the end of each row is a spring loaded door through which you can quickly slide out five or six nails, all in the proper nailing position. It takes a little time to get fully accustomed to the "stripper" but it is a great aid for speedy nailing. Do not overload the nail stripper. Too many nails jamb the sifting action and prevent the nails from dropping down into

Nail "stripper"
Figure 1-4

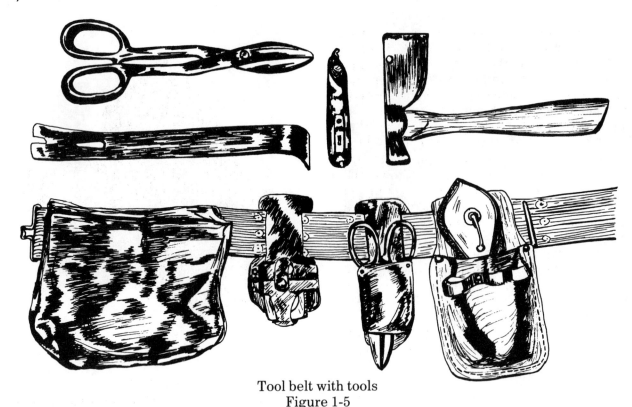

Tool belt with tools
Figure 1-5

the slots. A medium sized handful of nails is enough at one time. Wear your regular nail bag so that you will have more nails handy to drop into the stripper. A snug harness is necessary with this device so that you can easily jerk the nails free and keep the stripper slanted down toward the spring loaded doors. Occasionally a few nails will hang up at these doors. Shake the stripper to dislodge the problem nails. Though it may take time to get accustomed to using the stripper, you will find it worth the extra effort.

When a roofer goes up on the roof he usually knows what tools he will need and he should plan to take all the necessary tools with him. A tool belt will save time because you have all the necessary small tools within reach. In a simple tool belt, as in Figure 1-5, you can carry most of the small tools you need. The small pair of tin snips is for cutting metal edging. The tape measure is 16 feet long. The chalk line should be filled with either blue or yellow chalk before going up on the roof. The leather nail bag is a good idea because it allows you to carry more nails and get to them easier. Notice that the nail bag has been altered so that it will not hang down as far as a regular carpenter's nail bag. This is important because a roofer will either be sitting or squatting most of the time. Because of the sitting or squatting you do, you should wear a loose pair of work pants, preferably with a small leg pocket on the right leg. The pocket is handy for carrying your utility knife. This knife should have both hook blades and straight blades. Figure 1-5 also shows a big pair of tin snips for heavier cutting, a pry bar for removing nails, and a roofer's hatchet. These are the most common tools used in roofing.

Occasionally you will need other tools such as a hand saw, a key hole saw, a hammer, and two pairs of tin snips that are designed for cutting on the left and right. Also, you will quite often need a first aid kit. When the hook blade becomes dull you should change to a sharp blade and slip the blade back into the handle of the knife. Dull blades can be resharpened on an electric grinding wheel. After a little practice you should be able to sharpen these blades so that they will be sharper than they were when new. This will make your cutting much easier and also save you money. Straight blades are harder to sharpen and are much cheaper to purchase. You will probably decide to discard straight blades once they are dull.

### Roofer's Hatchet

The roofing hatchet in Figure 1-6 is actually a lather's hatchet. However, with a couple of small changes it makes the best roofing hatchet that you can use. Buy the lighter weight hatchet that you find in only a few lumber yards. Sawing off the end of the handle, as was done here, makes the balance much more even. When roofers drive nails they do not use an arm swing like a carpenter or a lather would. Instead, skilled roofers snap their wrist more because they drive short nails and must drive thousands of nails every day. Some roofers will lay as many as 30 squares of 3 tab shingles in one day. This means that they drive 9,600 nails in a day. With a long arm swing this would be nearly impossible. By sawing off the handle you can "pop" the nails in with a snap of the wrist. These hatchets come without any holes for the shingle gauge so you will have to drill your own. This is a little difficult because the blade is tempered. Drill slowly with a sharp bit. The serrated head of the hatchet will help keep the head from sliding off the nail and striking your finger, especially when the hatchet strikes a nail that has a lump of galvanizing material stuck to the top of the head. The blade is necessary to hold the gauge for strip shingles, 3-tab shingles and wood shingles. For wood shingles the blade is also used to cut and split the shingles. When you are laying T-locks, the blade is handy for opening the slots so that you can slide the "ear" portion into place.

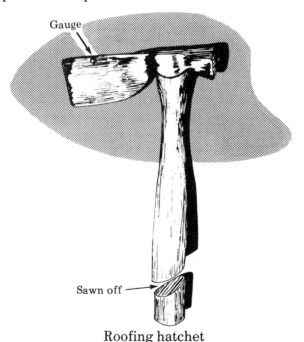

Roofing hatchet
Figure 1-6

Since the hatchet is the tool that you will use the most, it should feel comfortable in your hand. Many good roofers use heavy hatchets and unconsciously use only their arms to drive nails. If these roofers would try the lighter weight hatchets with the shortened handles they would wonder how they ever used the heavy hatchets. The right hatchet will help you drive more nails faster and easier.

Hip pad
Figure 1-7

## Hip Pads

In Figure 1-7 is one of the least expensive and most useful pieces of equipment the roofer will need. The hip pad illustrated was cut from the inner tube of a truck tire, but you can purchase a professionally manufactured hip pad that does the same job. The hip pad helps hold you on the roof, keeps your pants from wearing out and provides a little cushion between you and the rough shingle surface. It also insulates you from a hot roof. Sometimes these hip pads can be uncomfortable and hot, but they are essential. If you cut one of these from a truck tube, select an old tube that is thick, not dry rotted, and as big around as your leg. Cut off a tube section about 18 inches long. Then, cut out a strip about two inches wide down the inside of the tube, down to within about one inch of the bottom. The bottom will be left intact so that you can slide your leg through. At the top cut three or four slits for your belt. Weave your belt through these holes and through your belt loops so that the pad will stay in place. The natural curve of the inner tube matches the curves of your hip when sitting on the roof.

## Roof Jacks For Steep Roofs

One of the simplest and the most widely used roof jacks is the flat 2 x 4 jack. On the average house with 12 in 12 pitch or less, this type of scaffolding is easy to use and safe, providing that the 2 x 4 has a straight grain and is not over ten feet long. A board that is turned on edge as illustrated in Figure 1-8, is much stronger than a bigger board laid flat. Another

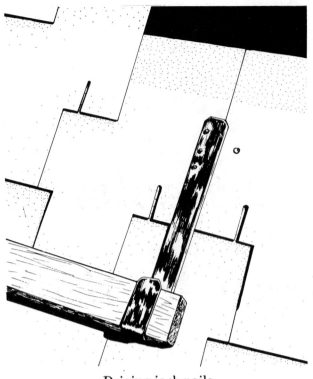

Driving jack nails
Figure 1-8

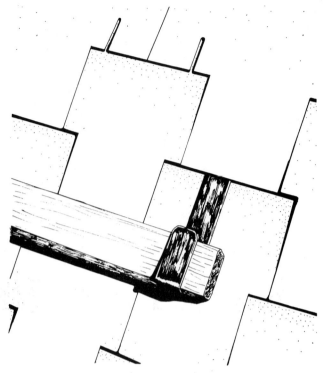

Positioning of the roof jack
Figure 1-9

popular roof jack is the solid 2 x 8 jack that allows you to slip in a 2 x 8 board. These jacks are built so that the board is perpendicular to the roof, thus allowing the roofer to walk on the flat part of the board. These jacks may not meet the requirements of the Occupational Safety and Health Act under some interpretations, but they are safe roof jacks when used correctly. Don't use a board that is too long. Handling a long heavy board may be unsafe on a steep roof.

When you nail down the roof jacks, always drive the nails where they will be covered later when another shingle is laid. The best way to remember this is to always drive the jack nails where you drive shingle nails in the shingle See Figure 1-8. The roof jack should be positioned as in Figure 1-9 so that you can easily lay a shingle over it. Always use long nails to anchor roof jacks and make sure that these nails hit something solid. When you finish with the roof you should start at the top and remove the jacks in the opposite order that you installed them. With the jacks illustrated you can remove the board and with your hatchet, tap upward on the bottom of the jack until the jack falls free. For a smoother roof it is a good idea to raise the tabs covering the jack nails and drive the nails down flush.

Roof jacks you use will vary depending upon what is available in your area and what is needed on your particular job. But don't try to get by with inadequate jacks. Select good straight grain lumber and test every board before using it. If you use boards over ten feet long, you should install an extra jack in the middle of each board. Remember that you will probably have the extra weight of a bundle of shingles on the board. Be sure to not overload any one scaffold board.

# Chapter 2
# Basics of Shingle Roofing

This book has been written for the journeyman roofer. However, with a little basic information, the beginner, homeowner or builder with little experience in roofing will be able to follow the explanations given here. This chapter has the basic information all roofers use in their work. Skip over anything that is elementary to you. Refer to the definitions at the end of the chapter as necessary. If you have been working as a roofer for some time you probably won't miss anything if you go on to Chapter 3.

### Shingle Types

The most widely used shingle is probably the 3-tab asphalt strip shingle. The next most common shingle in use depends on the area of the country where you work. Regardless of the type, all shingles provide excellent coverage for sloped roofs when applied correctly. While 3-tab shingles are called strip shingles, there are many new strip shingles that are not 3-tab shingles. There are 2-tab shingles, strip shingles with no cutouts and strip shingles with an overall shake appearance. 3-tabs have also been called square-butts and thick-butts because some manufacturers produced a 3-tab shingle that was thicker at the butt edge. Some manufacturers still make a thick butt shingle. This type shingle is tapered slightly, similar to a wood shingle. Figure 2-1 illustrates the shapes of common roof shingles.

Normally a good asphalt shingle roof uses a double coverage shingle. Most strip shingles and T-lock shingles are designed for double coverage. To be double coverage, the shingles must overlap each other by slightly more than half, thus providing two layers of shingles over the entire roof area. By overlapping more than half there is a small area that will be three layers thick. This is called the headlap. Most of the new shingles with a shake appearance are double coverage. A few provide triple coverage. Most of the less expensive shingles provide for only a single coverage. However, some of the most expensive shingles such as tile and aluminum use only single coverage. Wood shingles provide triple coverage. Shake shingles are double coverage when they are applied at the recommended exposures. See Chapter 11 for more details on wood shingles.

### Beginning A Roof

With new roofs the first step is to inspect the entire roof area for unnailed sheathing, weak spots and trouble areas. Notify the builder or the carpenter foreman of anything unusual so that it can be repaired. After this, the entire roof must be swept clean. All roofs (except wood shingles)

Roofers Handbook

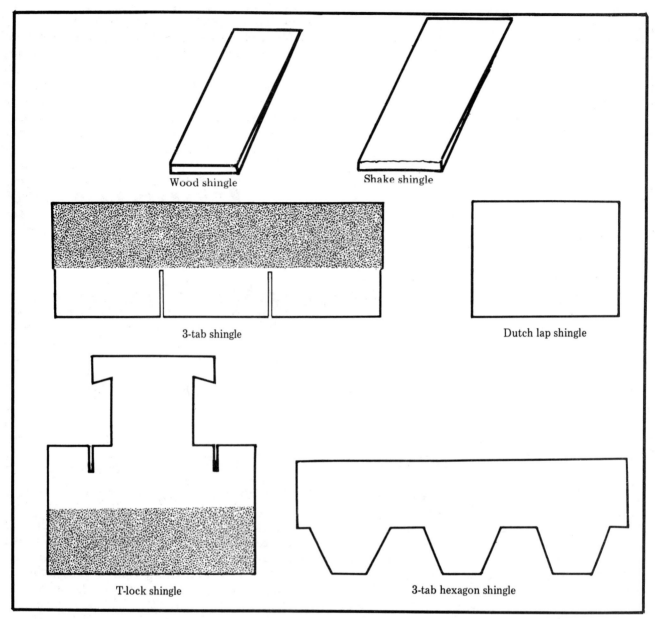

Shapes of common roof shingles
Figure 2-1

require a felt underlay, normally a single layer of 15 pound asphalt felt, perforated if available. Start the felt at the very bottom edge, staying as close to the edge as possible, and run the felt across the roof lengthwise. Use only enough nails to hold the felt smooth and in place until it is roofed. Load the shingles on the roof to help hold the felt in place. Store most of the shingle bundles on the upper two thirds of the roof area. This will leave you with enough room to start the roof. As you apply shingles, work up to the stored shingles, using up the nearest bundle as you proceed. If the roof is properly loaded you will not need to move any bundles out of your way.

On reroof jobs a felt underlay is not needed or required by the manufacturers. In most cases the felt would prevent you from using the butt-up method that is explained in Chapter 3. Make sure that the roof is clean and dry before you begin. The first step is to trim the old shingles back from the edge so that metal edging can be applied. Under no circumstances should metal edging be applied over the top of the new shingles as this will collect moisture and cause the lumber to rot. Proceed with the roof and clean up all trash and debris when you are finished. Inspect and clean the gutters if necessary as this might save you a ''call back'' later.

*The Basics Of Shingle Roofing*

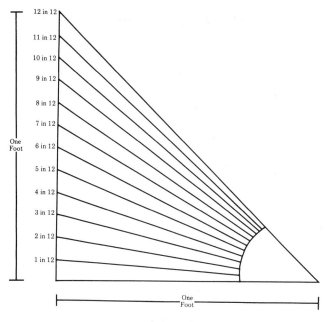

Determining roof pitch
Figure 2-2

### Roof Pitch

The pitch of the roof is the amount of rise in a level unit of run. For example, a 1 in 12 pitch roof will rise one inch in each 12 inches of horizontal distance. If the slope is 10 feet long, a 1 in 12 roof will rise 10 inches. A 12 in 12 pitch roof makes a 45 degree angle. See Figure 2-2. On a house with a 12 in 12 pitch, one slope of the roof is at a 90 degree angle to the other slope. A piece of paper held up by one corner will give you an idea of a 12 in 12 pitch roof. As the roof pitch goes up toward the vertical it becomes more like a wall. The International Conference of Building Officials, publishers of the Uniform Building Code which is followed by about 1,000 municipalities in the U.S., has ruled that a roof becomes a wall when the angle reaches 60 degrees or greater. A 60 degree angle is equivalent to approximately a 20 in 12 slope. That is a vertical rise of about 20 feet per horizontal run of 12 feet.

When you must determine the exact pitch of a roof before the shingles can be selected, the method shown in Figure 3-38 can be used. This could be important on low pitch roofs. If the pitch is under 2 in 12, shingles should not be used.

### The Importance Of Following Directions

Apparently the roofer that installed the T-lock roof in Figure 2-3 only glanced at the directions on the shingle wrapper. Some manufacturers suggest on the wrapper that the starter course be whole T-lock shingles turned with the locking tab up the slope. This entire roof was installed with the tab up the slope. However, the roofer did lock each shingle into place and then nailed each shingle properly with two nails. This is a classic example of not reading the directions. Don't allow the work you do to serve as an example of poor roofing practice. Understand what you are doing and work carefully. Chapter 4 has correct application procedures for T-lock shingles.

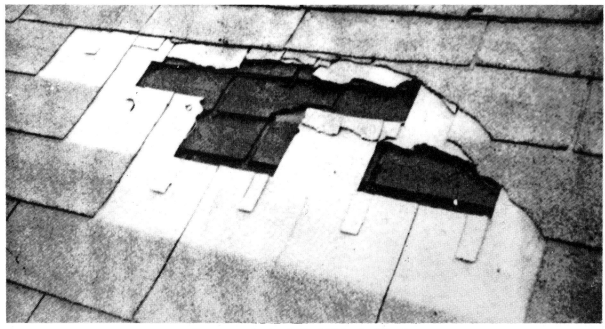

Improperly installed T-lock shingles
Figure 2-3

# Roofing Terms

Band Sticks. Boards used with wrapping bands on wood and shake shingles.

Blind Nailing. Placing nails in an area that will be covered by a shingle.

Blind Valley. A flat valley, one that cannot be normally seen from the ground.

Bond Lines. The alignment of the cut-outs on 3-tab or 2-tab shingles.

Bull. Name given to plastic cement in some areas.

Bull Paddle. The tool used to apply plastic cement, usually a narrow wood shingle.

Butt. The bottom edge or portion of a shingle.

Butt Nailing. The placing of nails in the butt portion of a shingle (face nailing).

Butt-Up. When reroofing, the alignment of the top edge of the new shingles with the butt edge of the old shingles.

Call Back. A return visit to a finished job because of a complaint.

Capping In. The application of the roofing felt to a roof deck (drying-in).

Counterflashing. The metal that laps down over the wall or step flashing.

Courses. Rows of shingles either taken vertically, diagonally or horizontally (also called runs).

Cricket. Wooden structure built at the high side of a chimney to divert water.

Cut-Outs. The slots or openings cut by the manufacturer in 3-tab, 2-tab and T-lock shingles.

Drift Down. Shingle course that angles down at one end.

Drying-In. The application of roofing felt to the roof deck (capping in).

Double-Coursed. The application of two layers of shingles per each row, normally with wood or shakes on sidewall application.

Ears. The small protruding portion of shingles that slips or locks into a slot on lock shingles.

Eave. The lower projecting portion of a roof.

Exposure. The amount of shingle left exposed to the weather.

Face Nailing. Nails placed in the exposed surface of a shingle.

Factory Edge. An uncut edge of a shingle.

Fascia. The wood trim covering the rafters and rafter ends.

Fishmouth. A buckle in a shingle.

Flashing. The metal used to connect shingles to a wall, vent, fireplace, etc.

Flue Cap. A waterproofing hood placed at the top of a chimney, usually fastened to the flue tile.

Freezeback. Thawing ice refreezing at the eave, forcing water back into the shingles.

Gauge. The act of measuring the exposure of shingles with a tool. Also the device used on the hatchet with which to gauge.

Granules. The rock like material used on the surface of shingles.

Hip Pad. A protective cover worn on the roofer's hip for protection and traction, usually made of rubber.

Laced. Woven or lapped back and forth.

Membrane. An asphalt saturated cloth used to strengthen strips of plastic cement when bonding and waterproofing a joint or edge.

Pattern. The design formed by the shingles when properly applied.

Pattern Lines. Normally the vertical alignment of the bond lines (cut outs) on a 3-tab or 2-tab shingle roof.

Patterns.

    4 Inch Pattern. On a 3-tab roof when the bond lines of adjacent rows are offset four inches.

    5 Inch Pattern. On a 3-tab roof when shingles are consistently offset to either the left or right forming slanted or diagonal bond lines five inches apart.

    6 Inch Pattern. On a 3-tab roof when the shingles are offset either left or right six inches leaving pattern lines six inches apart vertically.

    Half Pattern. When the shingles are offset one half tab. With 3-tabs six inches (a six inch pattern), with 2-tabs, 9 inches.

Pin Nails. Small galvanized nails (3d or 4d) used for face nailing on a ridge when needed.

Rake. The end of a gable slope, the outer edge of the first or last rafter.

Ribbon Course. A double layer of shingles used for visual distinction.

Run. A single row of shingles across the roof or a row of shingles in an application procedure, vertically or diagonally.

Saddle or Saddle Flashing. A water diverter used on the high side of a chimney, usually metal.

*The Basics Of Shingle Roofing*

Seal Cap. The layer of mortar at the top of a chimney.

Sealing A Vent. The application of plastic cement to a vent so as to make it watertight.

Serrated. Cross notching on the face of the hatchet head.

Shadow Course. A double layer of shingles used for visual distinction (ribbon course).

Single-Coursed. A single row of shingles when applying wood or shake shingles to a sidewall.

Slope. A continuous individual roof area. Also the angle or pitch of a roof slope.

Square. Unit of measure. One square is 100 square feet, equivalent to a 10'x10' area.

Squared Up. Having the shingles at a right angle to the bottom edge of the roof.

Stair-Step. The diagonal method of applying shingles.

Starter Course. The roofing material (roll roofing or shingles) applied to the bottom edge before applying the first row of shingles.

Straight-Up Method. The vertical application of 3-tab shingles.

Stripper. A metal device that separates and hangs the nails by the heads for easy handling. Worn strapped to the roofer's chest.

Tab. The individual exposed portion of a shingle.

Tail. The top end of a wood or shake shingle, the thin end.

Tie-In. The joining of shingles from separate shingle areas.

Toe Board. The scaffold board normally used only at the bottom edge of a roof.

Trim. The finish material (usually wood) that covers the rafters and rafter ends.

Tuck Pointing. The rebuilding or replacing of mortar in a stone or brick wall.

Underlay. The roofing felt applied directly over the roof deck.

Valley. The junction of two roof slopes that come together forming a V shape.

Valleys.

> Closed Valley. When the shingles join together in the valley (half lace and full lace).
>
> Full Lace Valley. When the shingles are woven or laced back and forth forming the valley.
>
> Half Lace Valley. When the shingles are lapped up on one slope and the shingles on the other slope are lapped back over these and then trimmed off evenly.
>
> Smooth Valley. A valley where the shingles on both slopes are trimmed back leaving a smooth water trough. The valley material is usually 90 pound roll roofing or flat galvanized metal.
>
> Woven Valley. A full lace valley.

Waste. The unusable shingle cut offs.

Water Guard. A turned up edge on valley metal or continuous wall flashing.

Water Trough. The cut back portion of a valley where rain water runs.

# Chapter 3
# Three Tab Shingles

There are several correct ways to apply three tab asphalt shingles. When selecting the method, pattern and type of shingle, you should consider first the roof surface itself. When reroofing, remember that three layers of roofing is about all that most roof surfaces will support. Where heavier weight shingles have been used or the rafters and sheathing are not adequate to support greater loads, only two roof coverings may be possible before the old shingles must be stripped off down to the sheathing. Also, some building codes dictate the maximum weights permissable.

Always select a pattern that will look good and be easy to install. Generally, the 5'' pattern (each shingle offset 5'' to the left or right of each shingle below) is used on hip roofs and the 6'' pattern is used on gable roofs. Some roofers, however, install a 5'' pattern on every house. On some roofs this would be the most difficult pattern because of the roof size and shape. You should consider whether the roof has a difficult tie-in, how many feet of side wall there are and whether you have a long narrow slope. A difficult tie-in is much easier with a 6'' pattern because it can be squared up as explained on the following pages. A large amount of side wall requires that every shingle be cut individually when you use a 5'' pattern. With a long narrow slope, about half of your time can be saved by using a 6'' pattern. With a 6'' pattern (sometimes called a ''half pattern'') you can use the ''straight up'' method and shingle right on up to the ridge much easier. Examine the roof carefully before you start and make sure you use the easiest and best pattern. When reroofing over wood shingles or asphalt shingles, always use the ''butt-up'' method to insure a smooth roof.

Sound craftsmanship requires that you plan the job carefully. The information on the following pages should help you select the pattern, application method and material that is correct for your job.

Six Inch Pattern Using The Stair Step
Method On A New Roof

Starting a roof with a six inch pattern and stairstepping it in is probably the most widely used roofing method. To start this pattern you need two vertical chalk lines six inches apart. Measure in from the rake 30'' and 36'' leaving the end of the tape measure sticking out over the edge ¾ inch to allow for the shingle overhang. This means that your lines are actually 29¼'' and 35¼'' from the edge of the wood or metal edging. First you should measure at the top near the ridge and place a nail at 30'' and 36''. Now measure at the bottom and mark the same. By hooking your chalk line on the nails at the top you can now chalk the two lines. Make sure that you hold the chalk line tight and always raise the line straight up so that you produce a straight line between the nails. Don't raise the line too high in a strong wind because the line will blow to one side and leave a crooked line.

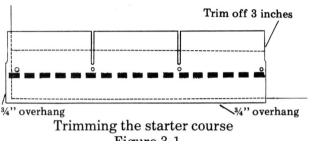

Trimming the starter course
Figure 3-1

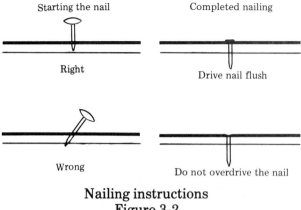

Nailing instructions
Figure 3-2

Now you are ready to install the starter course. Note Figure 3-1. First trim about three inches from the tabs. This eliminates a build-up of shingle thickness where the third shingle will overlap the second shingle. Lay the first starter shingle, with the cut edge up and the manufactured edge down, with the right end on the 36" line. Hook your index finger over the bottom edge of the shingle. With the tip of your finger on the wood of the eave, the edge of the shingle should come to the first joint in your finger. This will be about ¾ inch. Now check the overhang on the rake. If you have the 36" line in the right place you will have a ¾ inch overhang here also. Lay the next starter shingle to the side of the first and also on the 36" line. Adjust the bottom edge with your finger. Install the rest of the starter course in the same manner. Nail the starter course in place, driving the nails as shown in Figure 3-2.

Now you can start shingling with the number 1 shingle as in Figure 3-3. Lay the right edge of this shingle on the 30 inch line, keeping the bottom edge even with the starter course. Place

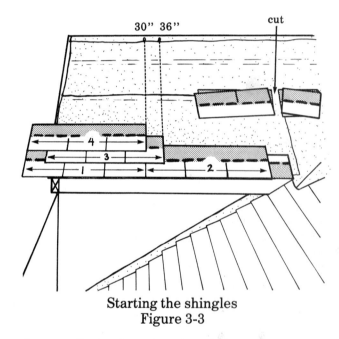

Starting the shingles
Figure 3-3

three nails in this shingle. Follow the nailing pattern in Figure 3-4. The number 2 shingle will also be placed on the 30 inch line alongside shingle number 1 with the bottom edge even with the starter shingle. Shingle number 3 will

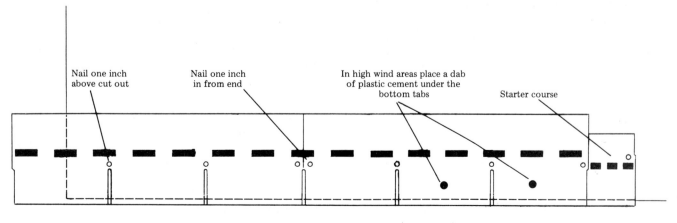

Nailing pattern
Figure 3-4

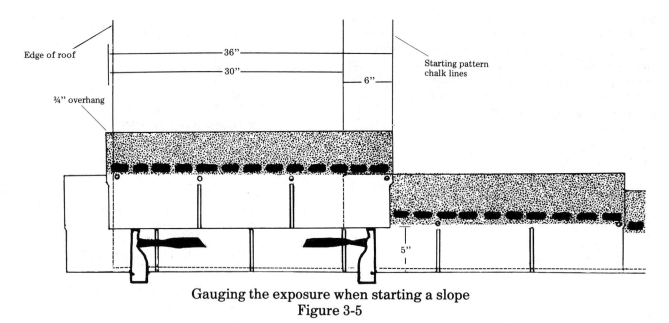

Gauging the exposure when starting a slope
Figure 3-5

be applied 5 inches up from shingle number 1 on the 36 inch line. Here is where you start gauging the shingles to keep the horizontal lines straight across the roof. See the gauging instructions in Figure 3-5. Shingle number 4 will be applied above shingle number 3 on the 30 inch line. Now cut two whole shingles so as to make two 2-tab and two 1-tab shingles as in Figure 3-3. Place one 2-tab shingle above shingle number 2 alongside the 36 inch line. Then lay another 2-tab above this on the 30 inch line. In Figure 3-6 this 2-tab shingle has been shaded. After laying the 2-tabs you must lay shingles number 5 and 6. The 1-tabs can then be applied nailing each tab with two nails as in Figure 3-7. This procedure will allow you to lay shingles number 7 and 8 to complete the stair-step diagonally across the roof. Now go back down to the bottom and start working your way back up to the top

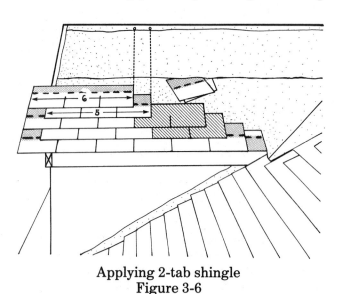

Applying 2-tab shingle
Figure 3-6

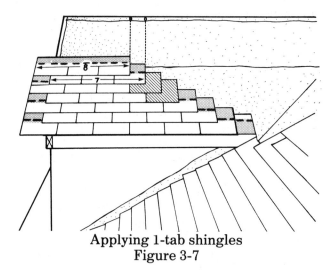

Applying 1-tab shingles
Figure 3-7

again. When you get up alongside shingle number 8 you will need to lay two more whole shingles on the chalk lines before you can lay more 2-tabs. This will be the same as laying shingles number 3 and 4. Then lay the two 2-tabs, shingles number 5 and 6, and the 1-tabs, and so on.

Usually on small slopes the "straight-up" method (laying entire vertical rows) is much faster and easier than stair stepping. On a larger slope once you get the shingles "stepped in" all the way to the ridge, the job goes very quickly. However, the bond lines tend to get out of line easier. If the bond lines must be as straight as possible, straight-up shingling will be the best method. The reason is that the shingles vary slightly in length from bundle to bundle. With the straight-up method you use shingles from the same bundle straight-up the roof instead of at an angle. The little slope in Figure 3-7 could be covered best using the straight-up method.

If shingles get out of line, the best way to straighten them is to simply reverse the step-in procedure until you have all the shingles straight up and down again. Start near the bottom alongside the third shingle with a 1-tab, then lay another 1-tab above it. Alongside of the fifth and sixth shingles lay the two 2-tabs. Now you can apply whole shingles until you get back up to the top of the slope. To continue the straightening up process start the 1-tabs again at the ninth and tenth shingle. Once you get all of the shingles straight-up in one row you can straighten the bond lines as explained later in this chapter.

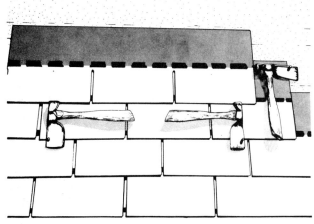

Gauging the second shingle
Figure 3-8

### Five Inch Pattern Using The Stair Step Method On A New Roof

This pattern gets its name from the 5 inch offset of each shingle. You don't need any vertical chalk lines to begin this pattern. Place your starter course as explained earlier. The first whole shingle after the starter must offset the starter at least 5 inches. You can measure this with your hatchet gauge. The second shingle is placed directly above the first shingle, leaving a 5 inch exposure and offsetting 5 inches to the left. Gauge this shingle in three different places, starting with the right end as in Figure 3-8. Hold the shingle by the top left corner with your left hand, positioning the shingle in about the correct location. Gauge the exposure on the right end and place one nail in that end. Now gauge the exposure on the left end and place one nail. Complete the nailing by nailing above the two bond lines. When you drive a nail, always be sure that the nail goes in straight. If the nail goes in at an angle it will probably move the shingle when you drive the nail down tight. Again refer to Figure 3-2 nailing instructions.

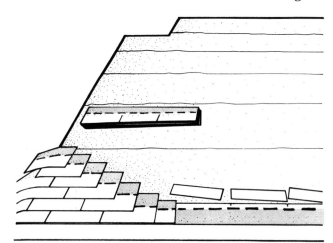

Applying the third through sixth shingle
Figure 3-9

Now you can lay the third shingle. Since the distance to the rake has been narrowed down, you must start laying shorter shingles so that the shingles do not droop over the edge and thus make the excess hard to trim off later. Simply cut two whole shingles so that you can lay two 2-tabs and then two 1-tabs. See Figure 3-9. This will complete one run in the stair-step method. Now you can start back down at the eave. Line the first shingle up even with the starter shingle. On the next shingle you will have to gauge only the right end. The left end butts with the other shingles. Continue shingling up and to your left in a stair-step fashion (Figure 3-10) until you get to the top of the slope. Then snap a chalk line and trim the excess shingle at the rake. If the shingles are left untrimmed for some time they will droop and become hard to trim off neatly.

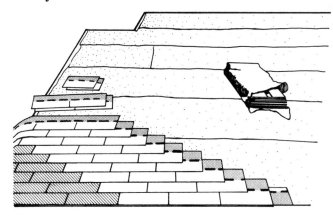

Shingling up in stairstep method
Figure 3-10

Measuring the shingles with the gauge on your hatchet may seem slow and awkward for a while but will become automatic and easy. The hatchet gauge is actually all that is needed to

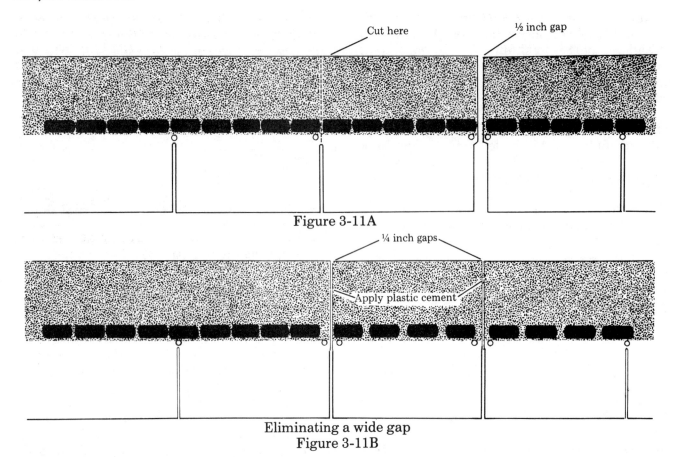

Figure 3-11A

Eliminating a wide gap
Figure 3-11B

keep the shingles straight. On a long slope the diagonal bond lines will start varying somewhat but these lines are less noticeable on a 5 inch pattern. If they are too crooked to continue shingling the rest of the slope, you can straighten them out using the following procedure. If you are stepped-in in one complete run, snap a chalk line along the top right hand corner of the shingles for the complete run. Since the shingles are crooked, there will be some shingles short of the line and some will probably extend past the line. When you place the chalk line, try to determine a point somewhere in between the extremes before snapping the line. If you start too short you will have to trim off excess shingle on every course all the way to the ridge. If you start too long you will probably leave a gap on many shingles. By snapping the line half way between the longest and the shortest shingle you will be able to straighten the bond lines easier and the joints will be less noticeable. When you start shingling at the bottom, always use the same hatchet gauge that you started with and regauge every shingle all the way to the top. If there is a gap where the new shingles join the other shingles you can adjust as in Figure 3-11A and B. If there is some excess to trim, it should be cut from the misaligned shingles rather than the shingles you are now laying. See Figure 3-12. Be sure that you use the same hatchet gauge to start a 5 inch pattern and to straighten out that 5 inch pattern. If more than one roofer is working on a slope, each roofer should stay in the same area. For example, if two roofers are working on one slope, the one that starts the slope should run the pattern all the way to the top. Meanwhile, the other roofer can be shingling across the bottom half of the roof. This will make it easier to keep the shingles straight vertically and horizontally. Even when this is done, the two roofers must have their gauges set as close together as possible. New shingler's hatchet gauge lengths vary between different brands. Also, there will be some difference between a

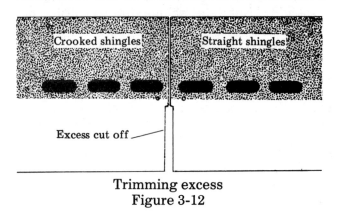

Trimming excess
Figure 3-12

*Three Tab Shingles*

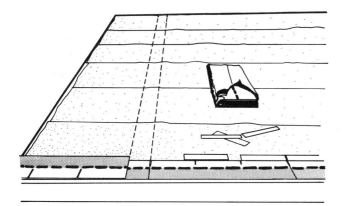

Starting the straight-up method
Figure 3-13

new hatchet and an old hatchet. To adjust a gauge that is set in a drilled hole, simply file down the side of the gauge that is short.

### Six Inch Pattern Using The Straight-Up Method On A New Roof

Shingling straight-up the roof makes it easier to keep the vertical bond lines straight. However, using this method it is a little harder to keep the horizontal lines straight across the roof. As always, it is important that you get started out straight. Start this method the same as you would start the stair step method. Allowing for your ¾" overhang, chalk two lines 30" and 36" from the rake as in Figure 3-13. Then apply the starter course and trim off about 3 inches of the tabs. See the starter course instructions in Figure 3-1. Since your starter course was probably set on the 36" line, your first whole shingle must be set on the 30" line. Now lay another shingle alongside the first whole shingle. The next shingle will be on the 36" line. The next shingle will go alongside of this. Now shingle on up a few courses in this

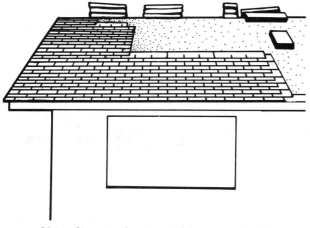

Shingling in the straight-up method
Figure 3-14

fashion as in Figure 3-14. Laying two courses, one shingle at a time, on each side of the chalk line as you shingle up the rake will help you keep the shingles straight. After you get two "runs" up the rake it is easier to apply just one run at a time.

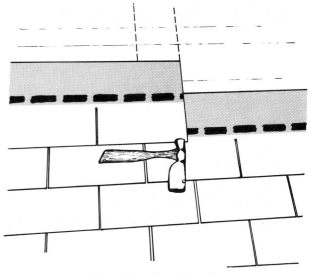

Gauging the shingle
Figure 3-15

Figure 3-15 shows the best way to gauge a shingle when you are starting on a chalk line. Gauge the right end first and then nail that end. Then gauge the left end and nail it. Now finish nailing the shingle as required. Once you are started out properly you should snap 2 or 3 horizontal chalk lines so that you can keep the shingles straight across the roof slope. These lines will be on the 15 pound felt underlay. When laid out correctly the top of one row of shingles will fall on the chalk line all the way across the roof. When you shingle up to the next horizontal chalk line, the top of that row of shingles should fall on the chalk line and be parallel with the lower chalk line. To lay these lines out you must measure very carefully from the bottom of the roof. Normally you measure from the bottom edge of the starter course. Since the distance to the top of the first course is 12 inches, you must add 5 inches for every course of shingles up the slope. So if you want a horizontal line every ten courses you must figure 12 inches for your first course and 5 inches times the remaining nine courses. For 10 courses you would figure 12 inches plus 45 inches or 57 inches. See Figure 3-16. This is where you chalk the first line. On the next line you won't have to allow for the starter so just figure 5 inches times the number of courses. These figures are accurate assuming that your hatchet gauge is an even 5 inches. If the gauge is off 1/10th of one

# Roofers Handbook

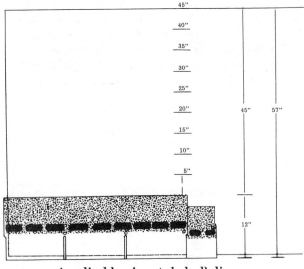

Applied horizontal chalk lines
Figure 3-16

inch, after you gauge only ten shingles you will already be off one inch. Keep this in mind when you measure the distance for your horizontal lines and adjust the figures as needed to fit the situation. When you shingle up to each chalk line you must make sure that the top of that row of shingles is positioned squarely on the line. Also, be sure to apply shingles that are of the same height because you will be gauging off of the bottom edge of these shingles for the next higher course. The lengths and widths of some shingles vary as much as ¼ of an inch from one bundle to the next. This is more than enough to get the shingles out of line if you are not careful.

Figure 3-14 shows a roof being applied using the straight-up method. This roof is being started so that two roofers can work together easily without crowding each other. They are using only one chalk line in the middle. The roofer below can shingle up to the line and then return to the bottom for another course. The roofer that is working above will always have a straight row to start out with each time. When there is only one person working on one slope it is better if he takes all of the shingles in one run all the way to the top of the slope. This is because some shingles vary in length about ¼ inch. Using these shingles in a straight-up method will make it easier to keep the bond lines straight.

When you are shingling straight-up you are offsetting each course six inches to the left or right. Shingles offset to the right should receive only three nails, leaving the right tab unnailed at the end. The shingles that offset to the left have to slide under an unnailed tab each time. After positioning the shingle and gauging the right end, start nailing at the right end. When you are ready to nail under the tab, shove your hatchet against the tab and raise it far enough to get your left hand under. Then hold the tab up with your thumb (Figure 3-17) and start the nail with your hatchet. While still holding the tab with your thumb, slide your fingers away from the nail so that you can finish nailing. Now you can nail the tab above that was left unnailed. Raising the overhanging tab and nailing the left end of the shingle may seem slow, but with a little practice it will become easy to raise the tab and nail it in one motion. Never cheat and leave out this nail. It is essential for a properly nailed and durable roof.

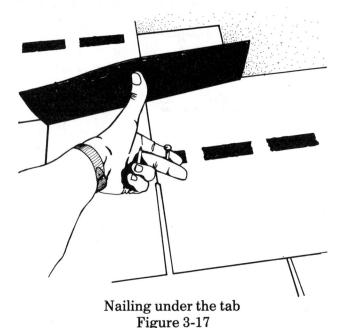

Nailing under the tab
Figure 3-17

### The Four Inch Pattern

The four inch pattern is probably the least used of the three different patterns for three tab strip shingles. See Figure 3-18. This is because it takes longer to lay out this pattern and get it started. This pattern is not as noticeable as the 5″ and 6″ patterns but may be more desirable in certain situations. The six inch pattern has a bond line that lines up every other course. The 5 inch pattern does the same except the bond lines are at a slight angle. The 4 inch pattern has a bond line that lines up every third course and the bond lines are only four inches apart. This is a very good pattern to recommend if all the roofs in the neighborhood are the same and the homeowner wants his roof to look different while still using the same shingle.

To apply this pattern you must start out with three chalk lines (Figure 3-19) instead of two as in the six inch pattern. Then use either the straight-up method or the stair step method. The main difference when you step-in the

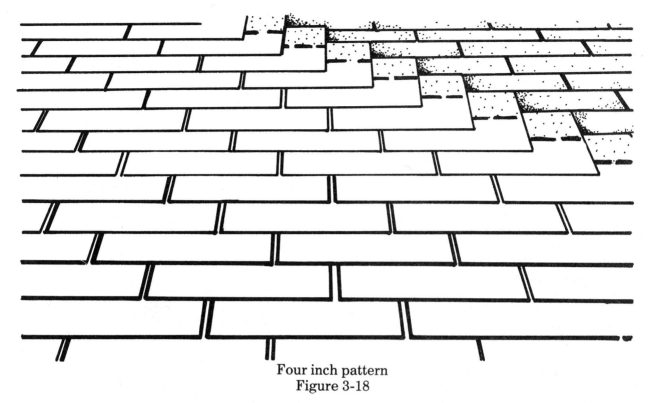

Four inch pattern
Figure 3-18

shingles is that you must use three 2-tabs instead of two and three 1-tabs instead of just two. The straight-up method should be used only on long, narrow slopes when using the four inch pattern.

### Straightening Vertical Bond Lines

When you notice that the bond lines are crooked, you should straighten the shingles out before proceding any further. Once the shingles are misaligned they can get worse. Straightening out is fairly simple using the following method. First, bring all the shingles in one row straight up and down as in Figure 3-20. If you have the roof stepped in, you can simply reverse the step in procedure as explained earlier in this chapter. Once you have your shingles straight up and down in one row you can see exactly how crooked the bond lines are. This is one of the advantages of shingling straight-up. You can easily detect when you are getting crooked and you are always ready to straighten the shingles.

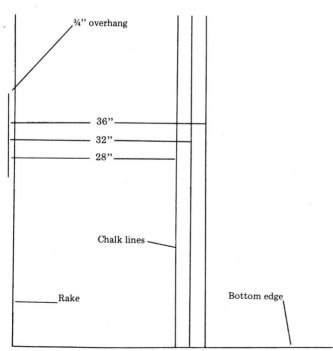

Chalk lines for four inch pattern
Figure 3-19

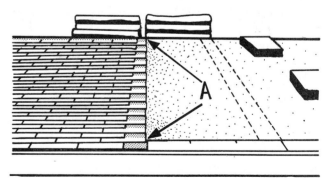

Straightening vertical bond lines
Figure 3-20

To straighten the bond lines you should first place a straight line or a chalk line without any chalk on it over the edge of the last course of shingles. See Figure 3-20 at point A. This establishes a measuring line for the straightening operation. Adjust the line so that it is not more than ½ inch away from the shortest shingles. This is necessary so that there is not a noticeable gap left when you begin shingling again. Now measure over from the line 30 inches and 36 inches at the top and bottom of the roof and chalk two more lines as in Figure 3-20. If you have the chalk lines in the right place you will have some excess shingle to trim off after you start shingling again. This trimming must be made on *the crooked shingles only* and not the new straight shingles that you will be laying. This is very important to the appearance of the two joints where you will be making the corrections or cuts. Note Figure 3-12. To make these cuts always use a hook blade knife. These blades are available at most lumber yards. Now use the 30 inch and 36 inch chalk lines to begin again as described earlier in this chapter. When you lay new shingles along side the crooked shingles, there should be no gap more than ½ inch. A gap less than ½ inch will be unnoticeable, especially since many shingles are slightly irregular. If for some reason there should be a gap of more than ½ inch, you should cut the whole last tab from the crooked shingle and slide it over ¼ inch or more to divide the distance. This will make the large gap much less noticeable. Note Figure 3-11. Doing this will leave a small gap under the tab above and you must apply a small amount of plastic cement to this gap to prevent any future leaks. Be sure that you don't use too much plastic cement. An excess will cause the plastic cement to ooze out onto the shingles.

This entire straightening procedure usually takes only a few minutes and is worthwhile providing that it is done neatly and accurately. When looking at the roof from the ground you should not be able to detect just where the straightening took place, except for the straighter bond lines. Remember that misaligned shingles reflect on the ability of the roofer as a professional craftsman.

### Straightening Up Horizontal Lines

On many jobs it is easy for the horizontal rows of shingles to become misaligned. When this happens you should see if one straight course can be laid above the crooked course. This can easily be checked by chalking a line at the top of the bond lines (cut outs) of the top course of the crooked shingles. If the shingles are too crooked to apply one straight course across the roof, you will need to use two separate courses. To determine just how crooked the shingles are, simply stretch a chalk line across the roof at the top of the bond lines. Now notice where the line is in relation to the bond lines across the roof. The chalk line should not be above the top of the bond lines. In some cases it will be, but it must never exceed ½ inch. Any time the chalk line falls above the bond lines you should apply plastic cement to the joints of the shingles under the tabs. So that the crooked shingles are not evident from the ground, don't allow the chalk line to come down on the bond lines more than two inches. Any time you exceed ½ inch above or 2 inches below the bond lines you should make the first chalk line curved so that you correct one half of the problem with the first new row. Then chalk a perfectly straight line. To stay straight, check the application of horizontal lines in Figure 3-16.

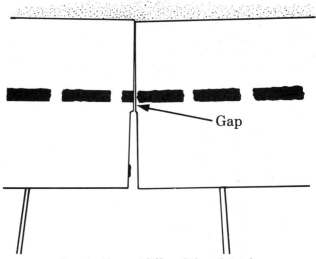

Gap in the middle of the shingle
Figure 3-21

You can notice that the shingles are getting a little crooked by the way that they are joining each other. If the joints leave a gap at the top the shingle course is probably going down a little. If the joints leave a gap at the middle, the course is going up. When the shingles are going up slightly, instead of leaving a gap in the middle, as in Figure 3-21, you should keep the shingles tight in the middle and let the top lap over the other shingle slightly. See Figure 3-22. This will keep the bond lines from getting crooked and will result in a tighter, more watertight joint.

### Eliminating The Overlap On The Starter Course

The importance of cutting 3 inches off the top of the starter course was mentioned at the beginning of this chapter. Note Figure 3-23. It doesn't look all that important but in Figure 3-24

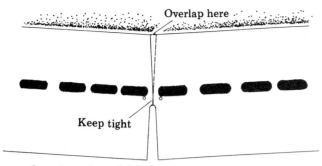

Overlapping shingles for straighter joints
Figure 3-22

you can easily see how this overlap causes a build-up of shingles in one area. This is an illustration of a tear-off in progress. There are two more 3-tab roofs under the roof surface. Each roof was gauged individually just like a new roof as though there was no thought to producing a smooth finished roof. The starter was not trimmed on any of the three roofs and the result was the hump as shown. Any time you do a reroofing job you should consider the overlap on the entire roof regardless of the type of roofing material used.

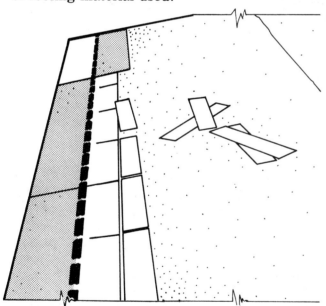

Cutting 3 inches off the starter shingle
Figure 3-23

### Offsetting The Overlap On Reroofs: "Butting-Up"

*Butting-up* is the term used to describe the offset of the overlap on a reroofing job. This offset is necessary when a 3-tab or strip shingle is installed over another 3-tab, strip shingle or wood shingle roof with an exposure of five inches or less. Some roofs have an exposure of four or four and one half inches and if you reroof over this you must butt-up to achieve a smooth

## Three Tab Shingles

Build-up caused by overlap of starter course
Figure 3-24

roof when finished. To install a butt-up reroof all that you have to do is position the new shingles up against the butts of the old shingles after you lay the starter and first course. The rest of the roof will be positioned two inches lower than the old roof. In some cases the new roof may require one extra row of shingles at the top. This depends upon how the old roof was finished out at the top. When butting-up, the only shingles that you need to gauge are the last two courses at the top. The horizontal lines will be exactly the same as the old roof, unless the new shingles vary slightly in height. It is normal for shingles to vary some and the small offsets across the roof will not be noticeable.

The reason for butting-up is illustrated in Figure 3-25A and B. Normally, on the average 3-tab roof there will be a two inch triple layer of shingles every three inches. This triple layer is what we call the overlap or the headlap. The three inches between the overlap is the double layer, usually called double coverage. So when you apply the new roof using the butt-up method you will be placing the overlap (triple layer) directly below the overlap of the old roof. When you reroof without using the butt-up method you will be placing the overlap directly on top of the old overlap. See Figure 3-25. The result will be a very uneven roof that will look even worse after the heat of the sun has softened and allowed the new shingles to lie down. Then the four layers between the overlap will sag as in Figure 3-26. This sagging is often the cause of roof leaks,

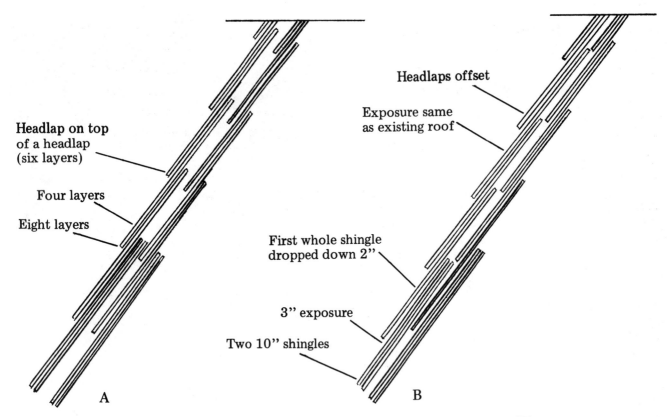

Reroofing without butting up (A) and with proper butting up (B)
Figure 3-25

usually after the roof is only two or three years old. This roof is also very vulnerable to hail damage because there is no support under the new shingles. If the butt-up method is used the new roof will be both smooth and tight. Offsetting the overlap is also important when reroofing with T-lock shingles over the same brand of T-locks and when placing roll roofing over roll roofing. On a 3-tab reroof, the only apparent difference, besides a smoother roof, will be a short course (three inches) at the very bottom.

If the old roof is very crooked horizontally or the exposure is more than five inches, you will need to either tear off the old roof or reroof with a lock shingle. A butt-up reroof under these circumstances will result in an uneven and inferior roof. A lock shingle has the ability to conceal more unevenness than a 3-tab shingle because of the basket weave appearance. If an old roof has an exposure less than five inches, you should use the butt-up method, but the roof will require more shingles than if a 5 inch exposure were used. With a 4 inch exposure on a 20 square house you will need 24 squares of shingles, about 1/5 more. With a 4½ inch exposure you must figure on using 22 squares instead of just 20, about 1/10 more. All of this should be explained to the homeowner because he may be getting bids from other roofing companies. If your competitors are not planning on butting-up, their bids could be somewhat less and you won't get the contract. If you explain

New roof that was applied without butting-up
Figure 3-26

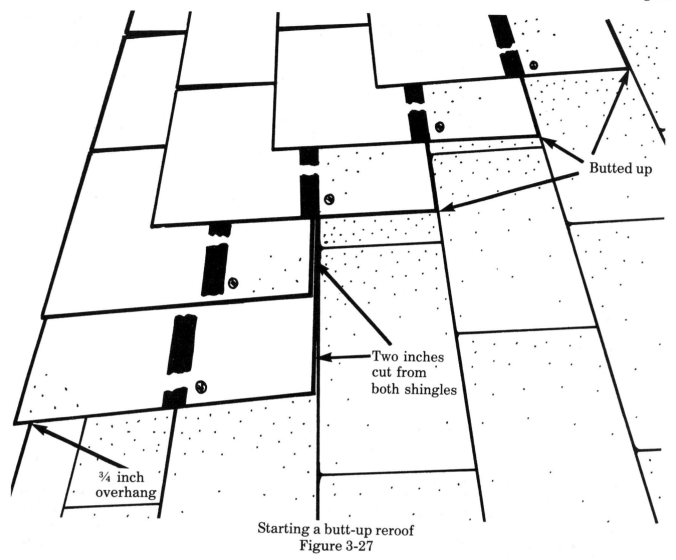

Starting a butt-up reroof
Figure 3-27

the advantages of a butt-up roof, the owner will probably respect your knowledge of roofing and your honesty and take this into account when selecting his contractor.

### Starting A Butt-Up Reroof

The correct way to start a butt-up reroof is the same no matter what pattern you use. First you must prepare the old roof. The old ridge cover should be removed, especially the hip ridge cover. If you are reroofing over wood shingles that have a metal ridge cover you will probably be better off if you leave it on. Usually when the metal ridge cover is removed a lot of wood shingles come off with it. Now trim back the old shingles around the edges and apply the metal edging. Regardless of what pattern you use, the following procedure will always apply when you reroof over a 3-tab, strip or wood shingle roof. The Asphalt Roofing Manufacturers Association recommends that the starter course be cut from a self sealing shingle, cutting off the tabs plus an additional two inches from the top, leaving a five inch portion of shingle. Place this strip at the bottom with the granules face up and the factory adhesive at the edge. Make sure that this strip does not lap upon the second course of shingles of the existing roof. The first shingle must have two inches cut off of the top and laid directly over the five inch starter course, keeping the bottom edges even and offsetting each other five inches. If for some reason the roof must be applied according to manufacturers recommendations, then this procedure should be used. However, realizing that roofers will take shortcuts whenever possible, we suggest the following method for starting a butt-up reroof. Figure 3-27 shows just how easy it is. The starter is an inverted three tab shingle with two inches trimmed off of the tabs. Most roofers use old shingles or shingles with damaged tabs for starter shingles. The first shingle has two inches trimmed off of the top. From here on all that you do is shove the new shingles up against

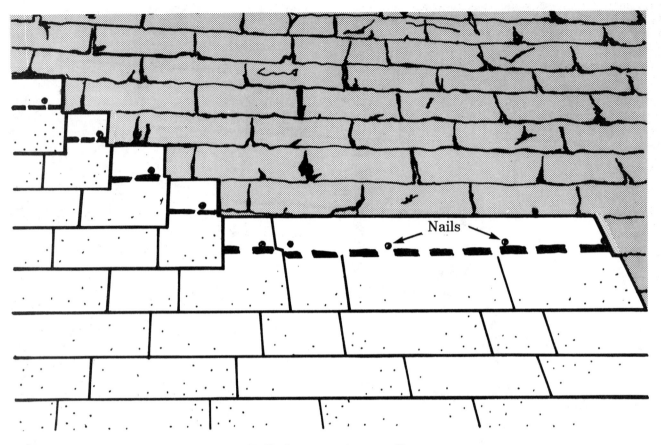

Nail placement in reroofing
Figure 3-28

the butts of the old shingles. When you get to the top you will need to gauge the last two shingle courses. If the old shingles are not square with the ridge, you can chalk a line for the last two courses so that the last shingle will be even with the ridge.

Note in Figure 3-28 that the nails are placed just above the factory adhesive on each shingle. This is very important when you reroof over rough asphalt shingles or wood shingles. The A.R.M.A. recommends that you always nail about 5-5/8 inch above the bottom of the shingle. On a butt-up reroof over wood or rough asphalt shingles you must nail high enough (7½ inches) so as to penetrate the tabs of the old shingles. See Figure 3-29A. This will permit the nails to serve an important secondary purpose, to pull down warped or curled shingles. The new shingles will lie down much better and form a better seal as a result of this nailing procedure. When you nail in the normal position you will always be nailing just below the butts of the old shingles. See Figure 3-29B. After the four nails have been driven properly there may still be one or two humps left because of the warped shingles. Go ahead and drive a nail in these humps so that the roof will be smooth. Be careful that you don't nail too close to the area where a bond line will fall on the next shingle above. If you do nail at a bond line you should cover the nail with a small amount of plastic cement so that there will be no future leaks as a result of this nail.

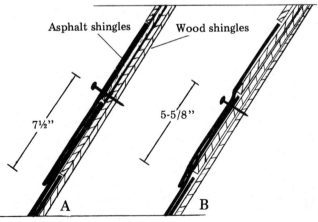

Proper nailing on a butt-up roof
Figure 3-29A & B

## Three Tab Shingles

Roof applied with butt-up method
Figure 3-30

### Illustrating A Butt-Up Reroof

Figure 3-26 shows the surface before the third roof was installed. You can tell by the uneven shingles that this roof was not butted-up when it was installed. Also note that the hail damage was in the low spots where the shingle had no support from below.

Figure 3-27 shows a clean smooth butt-up reroof. Note that the starter and the first course are both trimmed about two inches so that there will be no hump in the new roof. The three inch course at the bottom is normal and is hardly noticeable.

Figure 3-28 shows a strip shingle without any cutouts or bond lines being applied over a rough 3-tab shingle roof that probably should have been torn off. Note the nails are just above the factory adhesive. This will help make the new roof much smoother.

Figure 3-30 shows the roof in Figure 3-26 after it was roofed using the butt-up method. The new roof is definitely firmer, smoother and much easier to walk on. It will also be more resistant to hail damage.

Figure 3-31 and 3-32 show another roof before and after reroofing using the butt-up method. Note that the bottom edge of the old shingles are worn thin. If you cover up a roof like this, don't shove the new shingles too tight against the old shingles. You may have to gauge or estimate a couple of shingles now and then, but the result will be worth the extra effort.

Figure 3-33 shows a badly damaged roof. An old roof is seldom as badly damaged as this hailed out roof was. Normally you should tear off a roof like this. But here you can see how easily the butt-up method allows you to install a satisfactory roof.

### Starting In The Middle: Reroofing Straight-Up

When reroofing with 3-tabs and using the butt-up method, you can start shingling almost

Old damaged roof before reroofing
Figure 3-31

Roof after reroofing with butt-up method
Figure 3-32

31

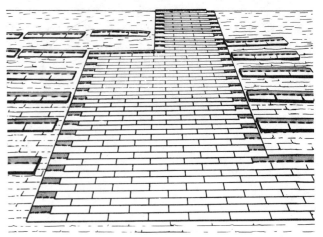

Butting-up with 3-tab shingles
Figure 3-34

Using butt-up method on damaged roof
Figure 3-33

anywhere that you need or want to. Note Figure 3-34. Half way between the left and right edges of the slope is usually the best point to start, especially on hip roofs. See Figure 3-35. This allows you to chalk the starting lines all the way to the top. On a large or long slope the bond lines should stay straighter if you start in the middle because you don't have as far to shingle until you reach the edge. Any time that you lay shingles continuously on one slope for 30 or 40 feet, the bond lines will tend to get a little crooked. This is unavoidable because of the allowable difference in the length of all brands of asphalt shingles. If you start in the middle of a 60 foot slope you will only be shingling 30 feet each way. With this method of shingling a right handed roofer can work on either side of the starter lines with ease. This method permits two roofers to work on the same slope without interfering with each other.

When starting in the middle you must determine exactly where you should chalk the two starting lines. You don't want the shingles to finish at the rake with a bond line on any course falling within two inches of the cut-off line. Usually you can look at the old roof, if it is a 3-tab shingle, and see how it was started and finished. Remember that you are using a 36 inch shingle and the starting chalk lines must be figured on a 3 foot increment. For example, 21 feet, 24 feet, 27 feet and so on. Don't forget to allow for the ¾ inch overhang. If trimming the shingles at the rake is difficult and there is enough shingle at the other end, you can allow an extra inch when you chalk the lines so that the trimming will be easier. When the temperature is warm, it is usually very difficult to trim off ½ inch or less. If you start in the middle of a slope and you don't have a square rake or old three tab shingles with which to measure and establish square vertical starting lines, you can use the "6, 8, 10" method of constructing a right angle. See Figure 3-36.

You don't always have to start shingling in the middle. If the ridge is level across the entire roof, you will want to start at the lowest point on the roof at the eave. If there is a wall at the right hand edge of a small slope, it would be better to start at the wall. Once this method of starting

Butting-up with 3-tab shingles
Figure 3-35

*Three Tab Shingles*

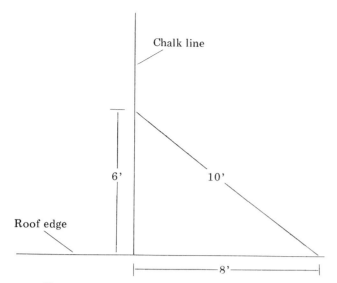

How to square up the pattern chalk lines
Figure 3-36

Old style gravel roof
Figure 3-37

anywhere that you want is mastered you will save a lot of time and effort.

### Gravel Roof Tear Off And Reroof

What was done in Figure 3-37 is what a lot of homeowners would like to do to their roof: tear it off and throw it away. In the 1950's many of the new homes had gravel roofs instead of composition shingles. For many reasons the homeowners have grown tired of these gravel roofs and most of them think that their roof is too flat to have a shingle roof. Two inch per foot pitch (2 in 12) is the minimum roof pitch that will accept asphalt shingles. For cedar shingles the minimum roof pitch is 3 inches per foot. Be sure of the roof pitch and never apply shingles on a slope that is less than is recommended here.

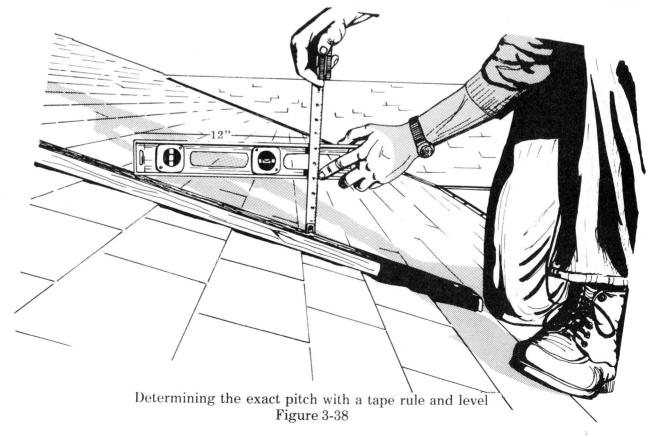

Determining the exact pitch with a tape rule and level
Figure 3-38

Roofers Handbook

Tearing off a gravel roof
Figure 3-39

Most gravel roofs are either 2 in 12 or 3 in 12 pitch. To determine the exact pitch see Figure 3-38. A 2 in 12 pitch roof should be roofed only with a 3-tab shingle or a strip shingle with no cutouts. A T-lock shingle is sometimes not as watertight on a low pitch roof such as this. An extra layer of 15 pound roofing felt or a heavier weight felt will not keep the water from getting under the shingles. Once this happens the nails will start rusting and the roofing felt will begin to rot. The best way to install a good watertight roof on a low pitch slope is to do as neat and smooth a job of application as possible. When applying the 15 pound underlay, make sure that it is completely free of buckles. If two layers of 15 pound felt are used, the problem of keeping the buckles out will only be compounded. The same applies for a heavier layer of felt such as 30 pound felt or a 43 pound base sheet. The air temperature and humidity are important factors in the smooth application of felt underlay. If the felt is left uncovered for a period of time and moisture collects on it, it should be permitted to dry thoroughly before applying shingles.

When tearing off a gravel roof (Figure 3-39), you must start at the top of a slope and work your way down. It will be a little hard getting started, but this way the gravel falls downward, staying on the old roof. The more gravel that you can keep off of the sheathing boards the easier it will be to clean up later. After all of the old roof has been removed, it must be swept clean, again starting at the top. Now you can see what repairs must be made. When this is completed you must pound down all nails that are not flush, again starting at the top. This pounding will loosen the gravel that is stuck between the sheathing boards. At the same time the loose gravel will bounce down the roof slope, allowing you to see what needs to be nailed down. Now the roof must be swept downward again. Be

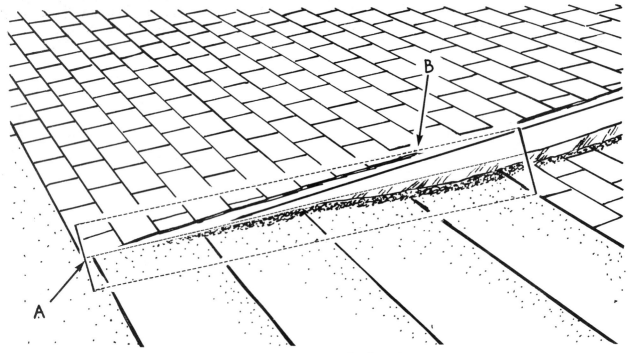
Flashing a low pitched roof
Figure 3-40

# Three Tab Shingles

Completed roof with asphalt shingles
Figure 3-41

thorough so you remove all the gravel. This is important because any gravel that is left will puncture the felt underlay and cause a hump in the shingles. This can result in a leak in the finished roof. Any gravel that is left between the sheathing boards or cracks will probably bounce up later when the shingles are nailed on. Then it is too late to remove the gravel.

Usually on a gravel tear off you must figure on replacing all of the vent flashings. On the roof in Figure 3-39 the roofer was very fortunate in that the old flashngs were in good condition and cleaned up easily. In Figure 3-40 you can see an offset where two sections of shingle roof tied into a low pitch roof. This flat roof had about a ½ inch per foot slope. It was roofed first using half lap roll roofing with a 15 pound felt underlay. The laps were sealed together with a cold application adhesive designed for this type of roofing. Where the flat roof came against the facia board, a solid strip of 12'' wide galvanized metal flashing was installed. The flashing was bent at 90 degrees for the length of the strip (7 feet) and was installed after the felt underlay but before the roll roofing, shingles and facia board were applied. From point A to point B the metal was bent over onto the other roof deck and nailed down. Plastic cement was troweled onto the metal flashing. When the roll roofing was applied, the edges that lapped onto the metal were pressed down firmly into the plastic cement. Later more plastic cement was applied to thoroughly seal this problem area. To make the joint last as long as the rest of the roof, a 4 inch strip of membrane was rubbed into the plastic cement and was then painted with aluminum paint. All of this may seem like a lot of extra work, but is necessary if the roof is to remain watertight for many years. Figure 3-41 shows the finished roof.

## Gravel Tear-Off

When a contractor estimates a job like this he should not give a firm estimate on the repairs needed on the sheathing. There is no way of estimating the amount of wood that must be replaced until after the roof is torn off. It is far better to estimate this type of work on a "cost plus" basis.

Remember that the main requirement for installing a good watertight roof on a low pitch roof is being neat and thorough. Be very careful around vents, chimneys and in valleys, always remembering that the water will run off very slowly.

# Chapter 4
# T-Lock Roofing Shingles

Although the different brands of T-lock shingles vary in size, the application method remains about the same. We will refer here to a double coverage shingle that is 19 inches long and 21½ inches wide. Brands vary somewhat, but when applied, all T-locks take on a similar basket weave appearance and make an excellent roof. The T-lock shingle also has the ability to cover up a rough roof that might otherwise have to be torn off.

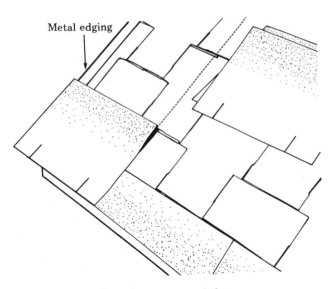

Starting out straight
Figure 4-1

The most important thing to remember about a T-lock roof is that you must start and stay straight. Figure 4-1 shows how to start straight. After the edges were trimmed back and the metal edging was applied, the starter was nailed on using the top portion of good T-lock shingles. One nail was placed at each end of the starter and a ¾ inch overhang was allowed. The first shingle was nailed on after the tabs were cut off. Beside this first shingle two more shingles were applied. This left room for the first whole shingles to be nailed on, as shown in Figure 4-2. Lock both of these shingles in place and line them up together before nailing them. They will be much easier to get straight this way. Now you are ready to apply a shingle at the edge. If the wind is not blowing or the sun is not too hot, you can lay a whole shingle and let half of it hang over the edge. If you must lay a half shingle you should cut it as described later in this chapter (Figure 4-14). Now you can lay two more shingles together as in Figure 4-3. Another row of shingles can now be brought up from the bottom. Each time you will be building the shingles higher. When you get to the top you should stop and cut the shingles that are hanging over the edge before they get too warm and droop down.

T-locks have definite vertical and diagonal pattern lines that are very apparent when

# T-lock Roofing Shingles

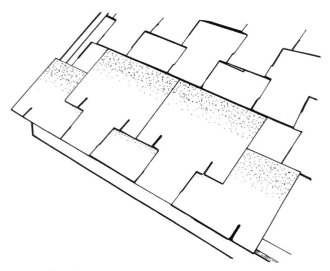

Laying the first two whole shingles
Figure 4-2

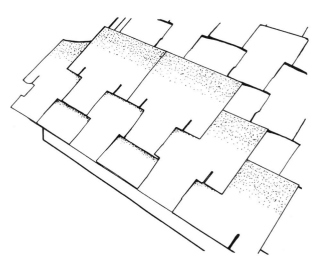

Proceding up with two more shingles
Figure 4-3

viewed from the ground. Consequently it is necessary to keep the shingles straight at all times. In Figure 4-1 the dotted line indicates where a chalk line could be used to keep the shingles straight when getting started. When you measure for this line, don't forget to allow for the ¾ inch overhang for both measurements, top and bottom. Some roofers do not start two shingles at a time as in Figure 4-2. Instead they try to line one shingle up with the rake, allowing for the ¾ inch overhang each time that they build one shingle higher. This method is good only if you are an experienced roofer and the rake is straight. If the rake is crooked the shingles up the rake will be crooked also. You should examine the rake before you start shingling. See if it is straight enough to follow. A vertical chalk line will help keep the shingles straighter in either case.

Proper nailing is important with T-locks just as any other shingle. For some reason more roofers nail in the wrong location when applying T-locks than with the other shingles. Figure 4-4 shows the nail location for most brands of T-locks. Two nails per shingle are required, approximately one inch in from the edge of the shingle and about one inch above the cut out. If you nail lower and closer to the cut out, the shingles will be harder to lock together. Some water will run under the tabs. Nailing in the location shown will keep the nails in the dry area.

When applying the shingle at the bottom, the shingle with the tab cut off, it is a good practice to leave about 1/4 to 3/8 inch gap every 15 to 20 feet. Keeping the first course tight sometimes causes the following shingles to be in a bind. The small gap simply leaves you with some room for adjustment.

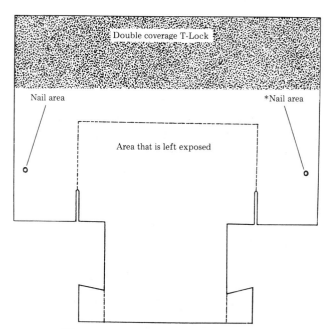

*Since an exact spot cannot be determined on each shingle, the general rule of nailing 1" to 2" above the slots and 1" to 2" from each side can be used. Only two nails per shingle.

Proper T-lock nailing
Figure 4-4

## Locking The Shingles

Installing or laying T-locks is a fairly simple procedure. By learning the proper method you will be able to lay more shingles per day, with greater accuracy and less effort. Note where the roofer is sitting in the illustrations. From there he can easily nail on 4 or 5 shingles before he has to move. Dont try to reach too many shingles from one position, but be sure that you don't get

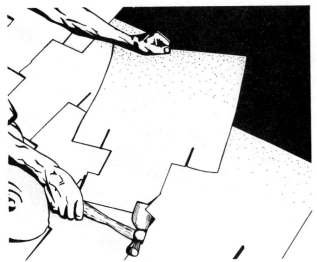

Using a hatchet blade to open the cutouts
Figure 4-5

Slipping the locks into position
Figure 4-6

into the habit of moving each time that you lay a shingle. Bad habits are hard to break and can cost you a lot of time. Use your hatchet blade to open up the cut-outs so you can slip the lock into position. Once you get accustomed to this method you will be able to lock in both "ears" in about 1½ seconds. See Figure 4-5 and 4-6. That may sound fast, but it is a reasonable expectation. If you have the shingles properly laid out before you, as illustrated in the following paragraphs, you should eventually be able to nail on a square of T-lock shingles in about 15 minutes. Some roofers can nail on a square of shingles in ten minutes, but this takes a lot of practice. Don't rush yourself.

Nailing with fingers held flat
Figure 4-7

Figure 4-7 illustrates the best way to hold your nails. Because of the serrated hatchet head and the short nails, it is very easy to miss the nail and strike your finger. Keeping your fingers flat on the roof can prevent repeated painful battering. The serrated hatchet is important here because some nails have a build-up of galvanizing material on the head. A smooth hammer or hatchet would slide off when you strike the nail. With a little practice you can get accustomed to nailing like this and your fingers will appreciate the difference. Note the extra nails in the roofer's hand.

### Tips On T-Locks

Instead of laying out your shingles on the roof deck in front of you, carry them on your leg as in Figure 4-8. This method may wear you out a little faster on a low pitch roof, but it will greatly increase your productivity. On steeper roofs, 5 in 12 or more, where you *and* your shingles will slide downhill easier, it will be necessary to carry the shingles on your right leg. Get into the habit of picking up one-third of a bundle each time. The extra weight on your right foot will give you much more traction. This will help hold you on the roof as your right foot usually doesn't have enough weight to get any traction. If your hip pad and right foot are not enough to hold you because the roof is too steep, you can try turning your left foot so that you will be using the sole. Always wear good high top boots with a soft crepe sole. This is very important. High top boots give your ankles needed support and the soft soles do not slide as easily as other soles. Be alert when walking on a roof. *Never* step on loose granules, the edge of a shingle or anything that is not nailed down. Any

*T-lock Roofing Shingles*

Shingles carried on the leg
Figure 4-8

Laying four courses at a time
Figure 4-9

time that the roof is steep enough (7 in 12 or more) a walk board or toe board should be used, at least at the bottom.

Note in Figures 4-8 and 4-9 that the roofer is laying only four courses across the roof. This method involves less movement. Therefor the roof can be shingled faster. Also, when the temperature is high, the shingles will get too warm to sit on. With this method you will be sitting on cool shingles until you must go back to the end and start a new set of courses.

that you are applying. This way you can keep the joints of the starter even with the other shingles and thus prevent leaks.

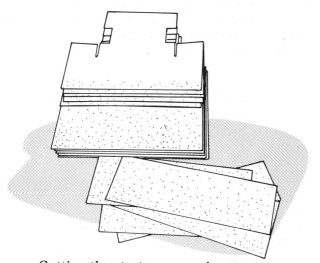

Cutting the starter course in one step
Figure 4-10

Figure 4-10 shows the correct way to cut the starter course in one easy step. Just take a whole bundle and cut about 9 inches off of the top of the shingles. The portion of the bundle not used as the starter can be used when you finish at the top. This narrow strip of shingle is all that is needed for a starter and it will complete the double coverage at the bottom. It is better to use tops of the same type of shingles

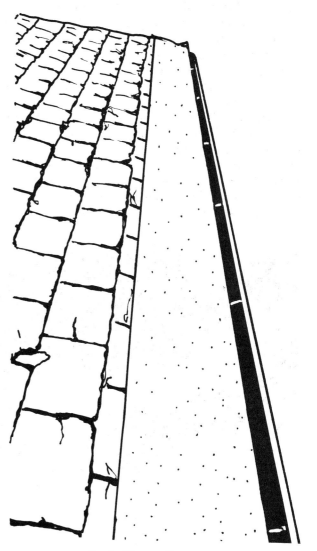

Installing roll starter
Figure 4-11

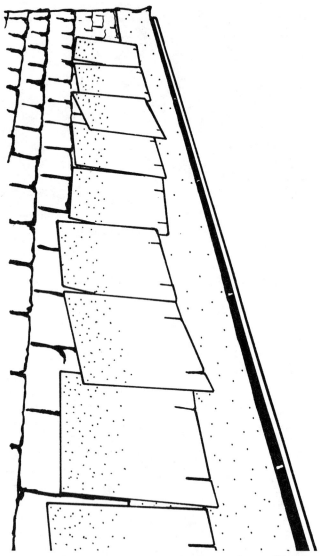

Laying first course in place for easier nailing
Figure 4-12

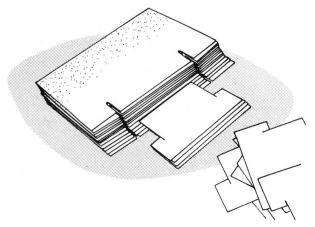

Cutting the tabs off in one operation
Figure 4-13

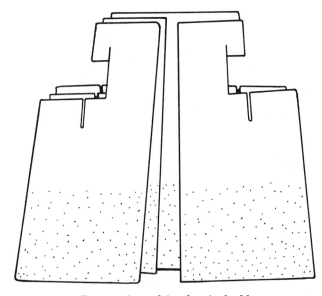

Precutting shingles in half
Figure 4-14

Another starter course that is popular is the factory made 9 inch starter roll. It comes in 36 foot lengths, four rolls to a unit. To separate these rolls, which are perforated, you must drop one end sharply against the ground. This should break them apart. The only problem that you might have when laying this starter roll occurs when the temperature is low. In cold weather you will not be able to simply roll the starter out and nail it on because it will stretch and buckle as it gets warm. If it is cold you should cut the whole roll into definite lengths. For example, if the shingles are 20 inches wide, cut the roll into 40 inch pieces. This is necessary to prevent buckling. The exact lengths are necessary to prevent the joints from falling directly under the lock portion of the T-lock shingle. While installing the starter, you should finish the slope that you are about to shingle, as in Figure 4-11.

In Figure 4-12 the roofer has the starter on and has laid out his first course so that he can reach them as he shingles across the bottom of the roof. The tabs were cut off of all the shingles in one operation, as in Figure 4-13. Some roofers nail on whole shingles and then come back and cut the tabs off later. This is much slower and harder. When you cut these tabs off you should lay them near the top of the roof because this is where you will need them later to finish shingling that slope.

Occasionally there will be a reason to precut your half shingles. Note Figure 4-14. Measure accurately and cut the shingles straight because this is a finish cut that will require no additional cutting later. The left over halves can sometimes be used at the other end of the roof. When shingling a steep roof, it is usually difficult for a right handed person to trim the shingles on the

*T-lock Roofing Shingles*

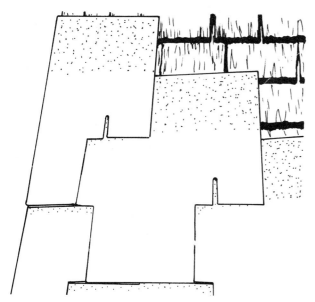

Using precut shingles on a left hand rake
Figure 4-15

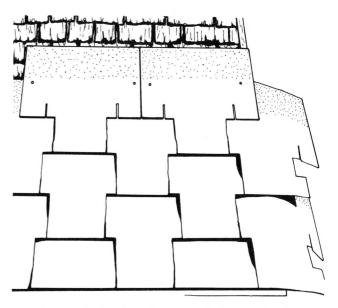

Using whole shingles on a right hand rake
Figure 4-16

left rake. This is another good time to use precut halves as in Figure 4-15. On the right rake of a steep slope it is very easy to trim the shingles so you can use whole shingles, as in Figure 4-16.

The only neat way to trim a rake is to chalk line it as in Figure 4-17 and then cut the shingles with a hook blade knife. You can make sure that you get the line in the right place by slipping your index finger under the shingles exactly where you want the ¾ inch overhang. Place a mark beside your finger and then check and mark the other end. Make a cut at the mark near the ridge so that you can drop one end of the chalk line in. This should hold one end while you stretch the line out and snap the line. Now trim the shingles.

### Straightening T-Locks

Sometimes on a T-lock roof, if the slope is a little out of square, you may get the shingles crooked. There is some slack in a T-lock shingle and if you use it correctly, the shingles can gradually be straightened out. In Figure 4-18 you can see where the roofer made a small adjustment in each shingle instead of one drastic adjustment. With T-locks you can't make all of the adjustment at once because two more shingles lock into each shingle. Figure 4-19 shows a roof that is in the process of getting crooked. In this case the shingles weren't pulled all the way up and were left a little way from the adjoining shingles. If they had been pulled up tight, the next row of shingles would not lock together.

### Offsetting The Overlap On T-Locks

To ensure a smooth reroof, use the same brand of T-locks and lower the shingles on the new roof about ½ inch. You can do this by starting as normal, putting the starter course on, then the first course without the tabs and then the first whole shingle. Now you can lower the second whole shingle about ½ inch, and the rest of the roof will follow. You can also lower the roof by starting with an extra ½ inch overhang at the bottom and then trimming it off later if necessary. The reason for lowering the roof is about the same as the reason for butting-up using 3-tab shingles. You will be offsetting the overlap or headlap. In Figure 4-20 you can see that lowering the new shingles ½

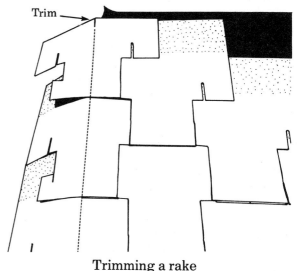

Trimming a rake
Figure 4-17

41

Keeping shingles straight by small adjustments
Figure 4-18

A roof starting to get crooked
Figure 4-19

inch keeps the tops from lapping up onto the bottom of the old tabs. This is where the overlap is. If you permit the tops to lap up on the old tabs, you will have a strip ½ inch wide on each shingle that will be 8 layers thick adjacent to a 4 layer area. There will be no support under the 4 layer area, making it more like to be hailed out. Lowering the new shingles will make the roof much smoother. It's simply another way to put on a better roof.

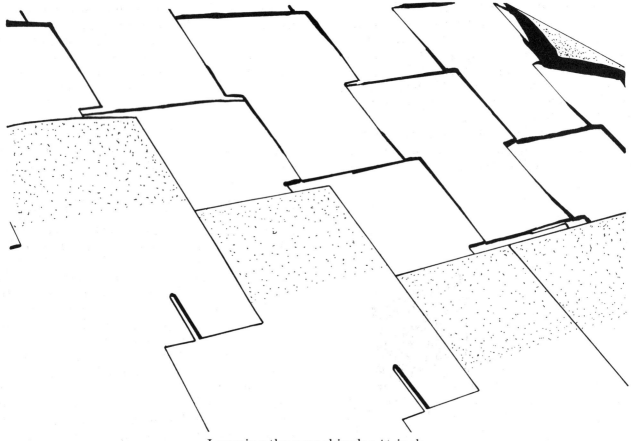

Lowering the new shingles ½ inch
Figure 4-20

# Chapter 5
# Shingle Tie-Ins

Not every roof is one long, rectangular, uninterrupted slope. In fact, relatively few roofs fall into this category. On nearly every shingle job there will be one slope where the roofer cannot continue shingling along the bottom edge. Then he must shingle up to the top of a dormer or valley keeping the shingles perfectly even. From that point he must transmit or project the pattern down to an exact spot at the bottom edge and start shingling again. Then he must bring the shingles back up to the top of the valley and tie-in to the other shingles so that the overall effect is one of smooth continuity. Good tie-ins result from maintaining good horizontal and vertical alignment. Good alignment is the result of planning and measuring carefully and following the layout you produce. There is no area of the roofer's trade where good professional craftsmanship is more evident.

### Tie-Ins For A New 3-Tab Roof Using A Six Inch Pattern

Many poor tie-ins are the result of not having enough of the slope shingled to know exactly what is needed. Figure 5-1 shows a simple tie-in where the bottom edges are even on both sides of the chimney. The roofer has already shingled the left side of the chimney. When he got to the first shingle that would go past the top of the flashing, he measured, chalked a line and then shingled above the chimney. To measure for this chalk line on the left side you must measure from the bottom of the shingles to the top of the course that will go past the flashing. On the right side you must let the end of the tape extend ¾ inch past the bottom edge and then measure up and place a mark at the same measurement as the top course. When you shingle along this line, line A, you must nail only at the top of the shingle so you can easily work up under it later when you complete the tie-in. The second shingle can be nailed normally.

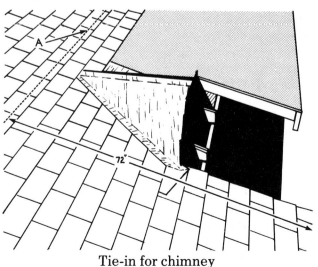

Tie-in for chimney
Figure 5-1

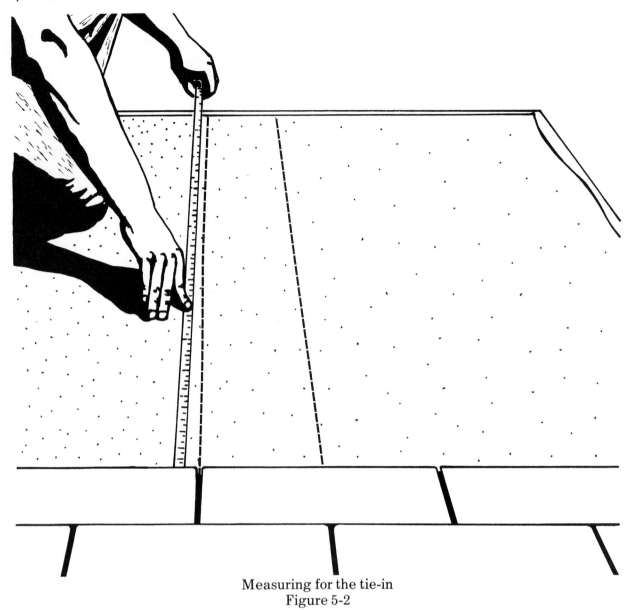

Measuring for the tie-in
Figure 5-2

In Figure 5-1 you can see that there are enough shingles installed to give the roofer something to work from. First, establish new pattern lines by lining up the chalk line with two bond lines on the shingles above the tie-in. Before snapping the chalk line, remove the chalk from the line that is on top of the shingles by snapping the line several times on a section of the roof that will be covered later. After chalking the two lines, one of these lines will connect to a cut out as in Figure 5-2. Now measure from the bottom edge of this shingle along side of this line to a point ¾ of an inch past the eave. This is important in determining the placement of the first shingle course. In this case the measurement was 60 inches. Since the 60 inch measurement results in an even number of courses, 12, he will start out at the bottom with a bond line on the same line that he measured alongside of, as in Figure 5-3. Note that the starter is positioned on the other line. Remember that with an even number of courses you start out on the extended bond line and with an odd number of courses you start out on the other line. This tie-in is now ready to be finished. The time required for layout was only a

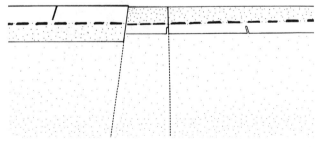

Starting out at the bottom
Figure 5-3

## Shingle Tie-ins

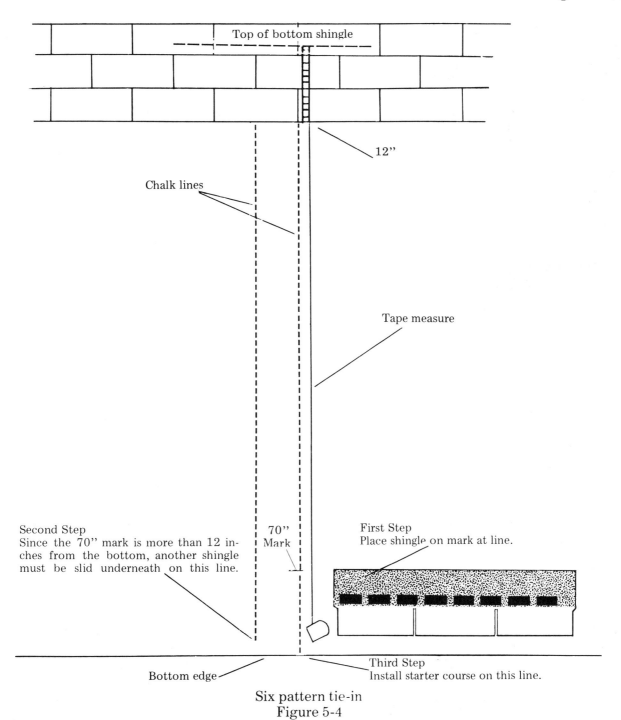

Six pattern tie-in
Figure 5-4

couple of minutes. When you shingle back up to the other shingles you must raise the tabs to complete the proper nailing.

### Tie-Ins With Uneven Eaves

If the bottom edges are uneven, you must measure differently. Shingle the roof as far as you can. Instead of measuring from the bottom of the shingle, as was done in Figure 5-2, slide the tape measure up twelve inches so that you will be measuring from the top of the shingle. See Figure 5-4. Now measure down to near the bottom edge and mark ten inch intervals. Place a mark on the felt paper at 70, 80, or 90 inches or whatever even number is closest to twelve inches from the bottom edge. See Figure 5-4. Assume the figure is 70 inches. Place a shingle at this mark with the end of the shingle on the chalk line that you measured alongside of. From this shingle you can easily determine what is needed to finish the tie-in. If this shingle does not reach the bottom edge, you will need to slide another shingle under it on the other line. Now slide your starter course under this shingle and

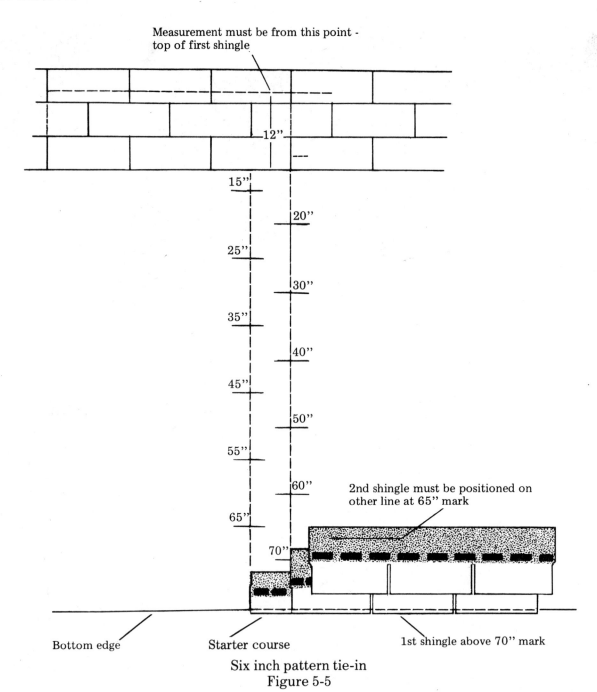

Six inch pattern tie-in
Figure 5-5

position it on the other line. Now that you can see exactly what is needed you can remove the shingles and, starting out with the starter course in the same place, start shingling.

If the shingle on the 70 inch line hangs over the bottom edge, you should bring it up and, allowing for the ¾ inch overhang, position it at the bottom edge. This will position the shingle above the 70 inch mark. You must then place the next shingle on a 65 inch mark. This is necessary so that the shingle courses work out evenly at the top where you tie-in to the other shingles. See Figure 5-5. As you shingle back up the roof, check the shingles to see if you will make the tie-in. The top corner of a shingle should fall on each mark, as in Figure 5-5. If you find that you are off a little, adjust the shingles before you get to the top. If you are off five inches, which is as far as you can miss the tie-in, make the adjustment equal over the remaining courses. If you have five courses left, adjust each course one inch.

When you have shingled up to near the other shingles, lay a few shingles on the roof in their right location to determine whether the tie-in is going to be successful. Don't try to stretch the exposure on the shingles when making adjustments. Instead, shorten the exposure. This method is called "stacking" the shingles and can save a poorly planned tie-in. The only thing that will be noticed from the ground will be a slight curve in the diagonal lines.

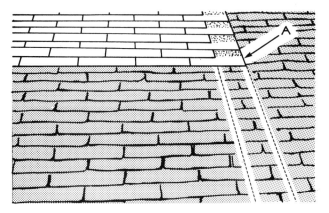

The straight-up tie-in
Figure 5-6

Returning to the other shingles
Figure 5-7

Butt-Up Tie-In Using A Six Inch Pattern

Successful tie-ins using the butt-up method are very easy. In Figure 5-6 the straight-up method of shingling is used. This requires the same tie-in procedure as the stair step method. With either method you need to chalk two lines six inches apart all the way to the bottom. These are your two pattern lines and they must be directly in line with the bond lines of the shingles above the tie-in. With the straight-up method you will be coming off of the end of the shingles as in Figure 5-6. With the stair step method you can chalk the two lines as in Figure 5-2.

Determine which line you need to start shingling from at the bottom edge by either of two methods. One method is to check the exact position of the bottom shingle at point A and note the old shingle under it. Now look down this same chalk line at the bottom of the roof. This should tell you at a glance exactly how to start out. This method will work only if the old roof follows a six inch pattern and if the shingles are straight. Another method that is more accurate is to place your right index finger at point A. Then raise the shingles up and place your left index finger on the other line one shingle down. Now place your right index finger on the right line, again one shingle below the last. Keep working your way on down to the bottom and place a mark where you need to start shingling. Now install your starter and start shingling. When you get back up to the other shingles, as in Figure 5-7, you must raise the tabs and complete the proper nailing.

T-Lock Tie-Ins On New And Old Roofs

A T-lock tie-in requires much more accuracy than other shingle tie-ins. If you miss the tie-in more than two inches you may have to tear the shingles off and start over again. Figure 5-8 shows a typical T-lock tie-in with a dormer in the middle of a slope. A right handed roofer usually starts in the lower left corner. To help him keep the shingles straight, he can put a chalk line (line A) at the rake, allowing for the ¾ inch overhang. Since the bottom edge in Figure 5-8 is short, you might want to measure down from the ridge at two different points and chalk another line (line B) to keep the shingles straight as you shingle up the valley. As in Figure 5-8, bring all of the shingles up even with the top of the dormer, with the top row high enough to go past the dormer. This row of shingles must be straight all the way in order to lay out for the tie-in. You may need to chalk line C before this top row is installed. If the eaves on both sides of the dormer are even, chalk line C can be measured from the eaves. If the eaves are not even, keep chalk line C even with the shingles that you have just installed. To determine just how straight these shingles are, you can measure down from the ridge and straighten chalk line C as required.

Now shingle across to the dormer, positioning the top of the shingles on chalk line C. After you reach the top of the dormer, continue positioning this row on the line until you are above the bottom edge as in Figure 5-8 and 5-9, shingle A. Place one nail in the very top edge of

Roofers Handbook

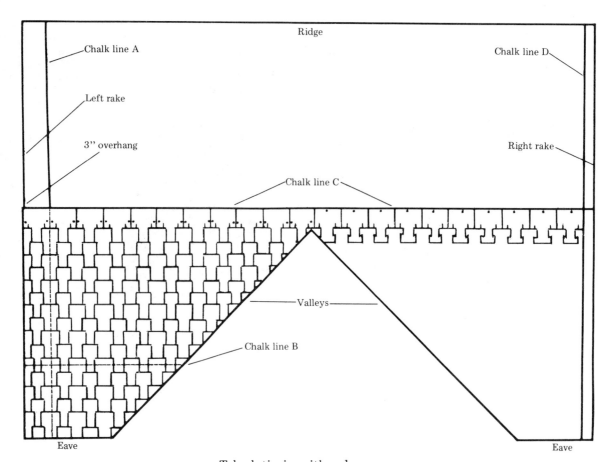

T-lock tie-in with a dormer
Figure 5-8

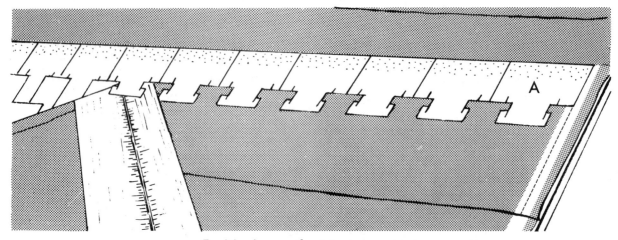

Positioning on the temporary row
Figure 5-9

this row of shingles because this is a temporary row of shingles. Now measure the distance from the edge of shingle A (Figure 5-9) to the right rake. In the illustration the distance was eight inches. Now measure this distance at the top and the bottom and apply chalk line D. On some roofs you may not have a rake from which to measure for chalk line D. Then you can lay out a square perpendicular line with a tape measure as in Figure 3-38.

Now you are ready for what is referred to as "backing down". This is the method used to determine the position of the starting course on the bottom edge. Backing down is merely simulating the location of the shingle courses on the finished roof. In Figure 5-10 you can see how

*Shingle Tie-ins*

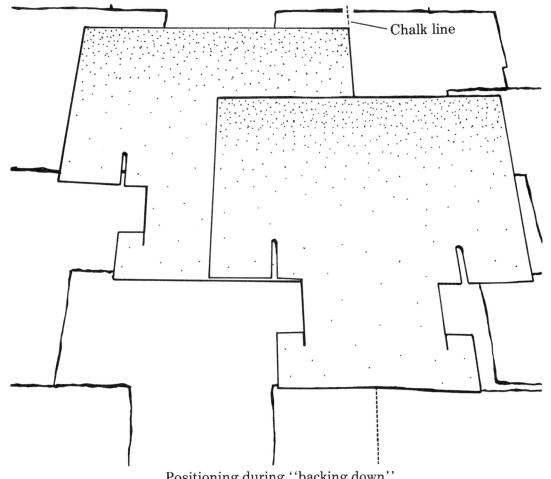

Positioning during "backing down"
Figure 5-10

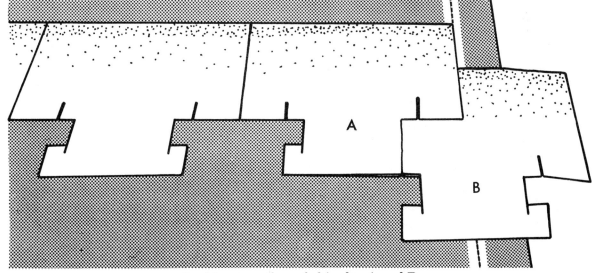

Locking together of shingles A and B
Figure 5-11

the shingles are lined up when "backing down". In Figure 5-11 shingle B is locked into shingle A to illustrate the relative position of the two shingles. In Figure 5-12 shingle B has been placed on top of shingle A to start the back down procedure. Now, while holding down firmly on shingle B, place another shingle in the position of shingle C. Then carefully slide shingle B out from underneath C and replace B in the position of shingle D in Figure 5-13. Then replace

49

Roofers Handbook

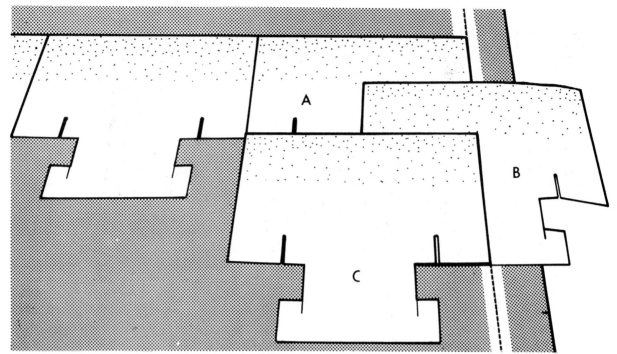

Starting the "backing down" procedure
Figure 5-12

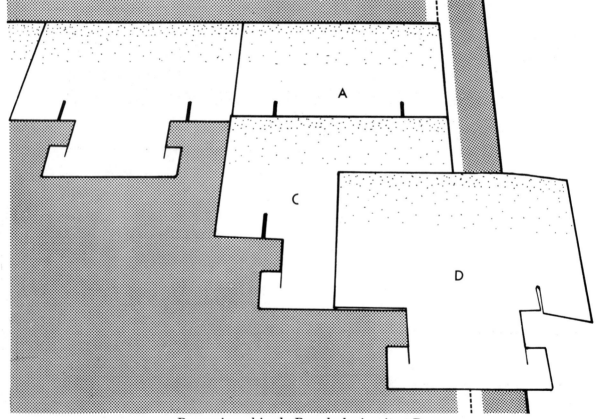

Removing shingle B and placing it at D
Figure 5-13

shingle C in the position of shingle E in Figure 5-14. Two shingles are all that is needed to back down, regardless of how far you must go. Line every other shingle up with the chalk line D in Figure 5-8 as you back down. With each step make sure that you don't accidently move one of the shingles. This could cause you to miss the tie-in. Continue backing down with the two

*Shingle Tie-ins*

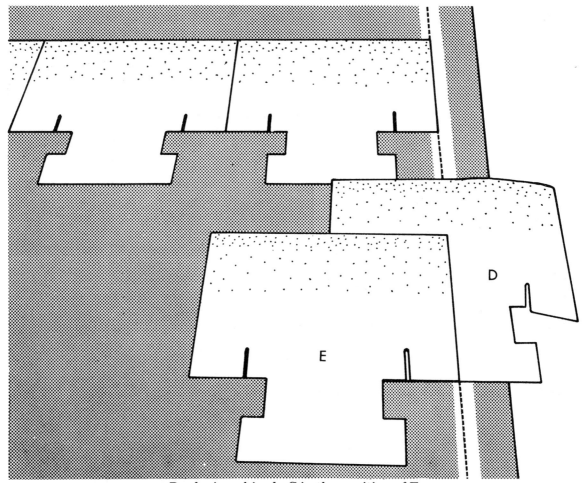
Replacing shingle C in the position of E
Figure 5-14

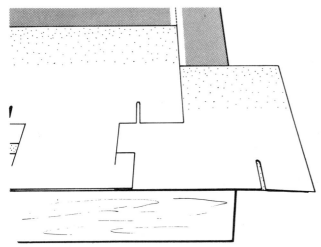
The last shingle in the "back down" procedure
Figure 5-15

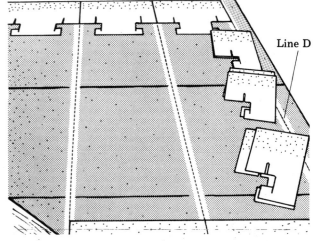
Using chalk lines to keep shingles in line
Figure 5-16

shingles until you reach the bottom edge. In Figure 5-15 the whole shingle was the last shingle used in the back down. If the bottom eaves are even, this shingle should overhang the edge ¾ inch. If the bottom edges are uneven, place two nails in the top of this shingle and from here you can see what is needed. You will need to let the shingles hang over the bottom edge. After the starter and the first course is applied you will need to trim the shingles to within ¾ of an inch of the edge.

Now you are ready to start shingling back up

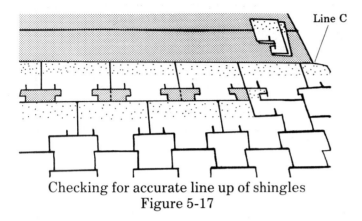

Checking for accurate line up of shingles
Figure 5-17

to the other shingles and complete the tie-in. This is where most tie-ins are missed because it is easy to get the shingles crooked when shingling up the valley. In Figure 5-16 you can see several vertical chalk lines. These lines are used to keep the shingles in line with the shingles above. Just measure over from line D the width of a shingle for each line you want.

As you shingle back up to the other shingles you must keep the shingles level with line C. In **Figure 5-17** the roofer has stopped short of the shingles on line C to see how the shingles are running. With two courses to go he can make any small adjustment that may be necessary. In **Figure 5-18** the shingles were installed up to shingle A to show the shingles that were backed down. Note that B and C are in the same position as B and C were in Figure 5-12.

There shouldn't be too much adjusting to the shingles when you approach the shingles on line C. If the shingles are a little too high you can lower each remaining course a little to make the tie-in. If the shingles are a little too low, simply pull the remaining two courses of shingles up tighter until you make the tie-in. If all steps of the tie-in were taken carefully you shouldn't be off very much. The shingles on line C can be removed as soon as you see that they are not needed any more. If they were nailed high

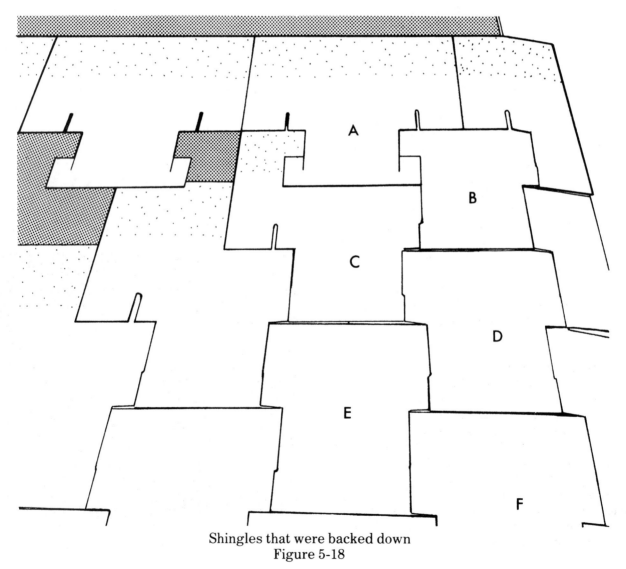

Shingles that were backed down
Figure 5-18

## Shingle Tie-ins

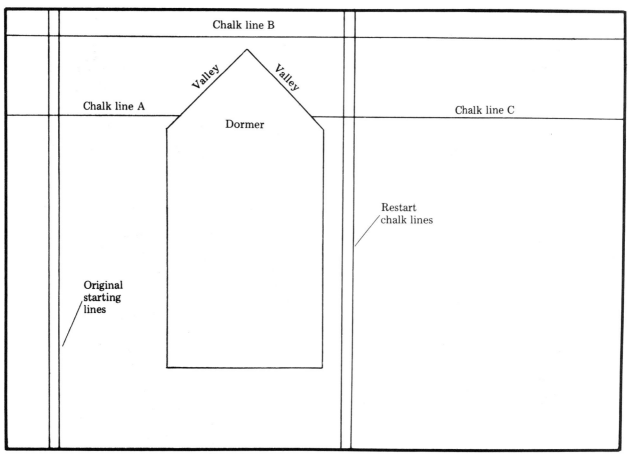

Keeping square around a dormer
Figure 5-19

enough you should be able to pull the shingles away from the nails and then drive the nails down flush with the roof deck.

This method of making tie-ins with T-locks is very reliable providing that the roofer doesn't get in too much of a hurry. When shingling back up to the other shingles you should make sure that the shingles are locked together evenly. When shingling in the area of the valley the shingles will have a tendency to drift down. You may need to pull these shingles a little tighter. This happens sometimes even to highly experienced roofers.

### Shingling Around A Dormer With 3-Tabs On A Six Inch Pattern

This is a tie-in that can be difficult if the shingles are not kept straight. In Figure 5-19 the dormer is positioned near the middle of a slope, leaving room for shingles to be applied all the way around it. You can start out shingling in the lower left corner as usual. The six inch pattern might be best on a roof as in Figure 5-19 because this pattern is easier to square up for tie-ins. First shingle the left side and the bottom of the dormer. After you get up to the bottom of the valley and start shingling into it, you may need to measure down from the ridge and chalk a line near the same location as line A in Figure 5-19. Then continue shingling until you reach the top of the valley. Before laying the first course that will go past the top of the dormer,

Laying shingles to the ridge and past the dormer
Figure 5-20

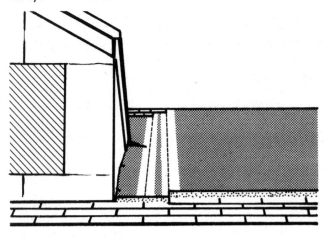

Chalking restart pattern lines
Figure 5-21

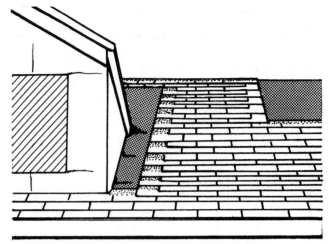

Shingles installed at new restart lines
Figure 5-22

Straight-up method with 6 inch pattern
Figure 5-23

measure and chalk a line even with the ridge all the way across the roof, the same as line B in Figure 5-19. Now lay the remaining shingles on up to the ridge and past the dormer as in Figure 5-20. Now chalk two new restart pattern lines (Figure 5-21) from the shingles at the bottom of the dormer to the shingles at the top. Go ahead now and shingle this area up to about the bottom of the valley. Again chalk a horizontal line (line C) to make sure that the shingles stay straight. Now you can shingle on up and make the tie-in.

In Figures 5-22 and 5-23 you can see that one course of shingles was installed up the new restart lines and that the shingles are tied together nicely. The remaining portion was shingled up to line C. Also note that by using the straight-up method with a six inch pattern it was easy to square up the roof and complete the tie-in without any noticeable adjustments. The proper use of the chalk lines made the whole operation much easier. While it looks like the layout was time consuming, no single chalk line took as much as a minute. Also, with the straight-up method all attention could be focused on shingling and completing the tie-in, leaving the metal flashing and sidewalls for later when it would be more convenient. Then the roofer could measure the two different lengths of shingles required and cut what he needed. Later he could devote all of his attention to the sidewalls and flashing. Being organized and using the proper methods make the whole job easier and faster.

Shingling Around A Chimney

New pattern chalk lines should be used when going around any object so that you maintain straight vertical bond lines. Note Figure 5-24. You will probably have misaligned shingles if you try to guess at the proper position or use the factory applied six inch notch at the top of each shingle as a gauge mark. This may be hardly noticeable on the roof but it usually shows up very clearly from the ground.

A Difficult T-lock Tie-In (Figure 5-25)

The standard T-lock tie-in which was explained previously will cover most situations. However, ocassionally you will have a tie-in where you can't chalk the lines that are normally required. In the other T-lock tie-in we explained how to back down the shingles to determine the position of the first shingle at the bottom. With this more difficult tie-in the roofer has to lay out one row of shingles across the slope, then back the shingles straight down and then back the shingles down at an angle until he reaches the bottom. All the shingles that are used in this backdown procedure are left tacked in position so they can be used as a guide when shingles are applied back up the slope.

Maintaining a straight line with new chalk lines
Figure 5-24

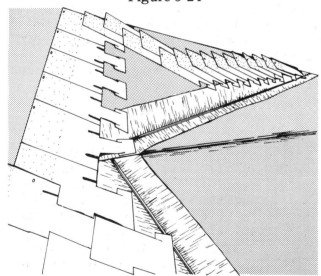

Using "back down" technique
Figure 5-25

To back down, start out by chalking the horizontal line C as was explained in Figure 5-8. Since the roofer here didn't have a rake to measure from for chalk line D and the distance that he had to go was short, he backed straight down using only the shingles to keep straight. This was accomplished by using a notch cut in the top center of each shingle. To cut this notch, just line up a half bundle of shingles, measure to determine the exact center, and then cut with the blade of a roofing hatchet. When backing down a short distance you can align the bottom of the shingles as shown in Figure 5-10 and align the top with the notch even with the shingle above. In Figure 5-25 you can see that the roofer backed straight down using six shingles which were tacked in place temporarily. Then the roofer started backing down at an angle toward the bottom of the slope, using the notch at the top of the shingles to keep straight. These shingles he also tacked temporarily in place. Upon reaching the bottom he simply nailed on the last shingle and then started shingling back up the slope, using the shingles that were tacked in place as a guide to keep straight.

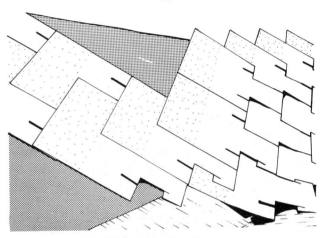

Shingling back up
Figure 5-26

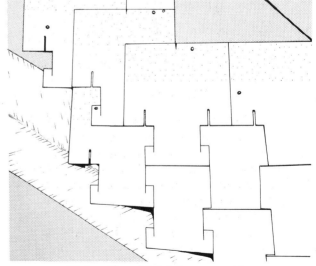

Using the shingles as a guide
Figure 5-27

In Figure 5-26 the shingles have been brought up to where the roofer started backing down at an angle. Figure 5-27 shows plainly how to use the shingles as a guide. After checking

55

Stopping to measure for straightness
Figure 5-28

Completed slope before trimming
Figure 5-29

and adjusting as necessary, slip the shingle out from underneath and nail the shingle that you are laying. As you work your way back up to the shingles on line C, you should stop two courses short, as in Figure 5-28, and measure to determine how the shingles compare with line C. On a narrow slope such as this it is easy to get one end higher or lower than the other end. Make any necessary adjustment so the tie-in will be neat and attractive. Figure 5-29 shows the completed slope before it is trimmed out.

It does take about five extra minutes to lay out for this tie-in. But how many hours are wasted if the homeowner or general contractor doesn't approve of a botched job. Any tie-in on any roof can be made successfully if a little effort is used in planning and layout.

### The Five Inch Pattern Tie-In Using 3-Tab Shingles

Many different methods are used in making the five inch pattern tie-in. However the method in Figure 5-30 is probably the most accurate. To start the tie-in, shingle as much of the slope as possible so that you will have a good guide for measuring and laying out chalk lines. Shingle up the left side of the dormer. After measuring on both sides of the dormer, chalk a line all the way across the roof. Now continue shingling above the dormer, but be sure to nail at the top of the shingles that are on the chalk line above the tie-in area. After you have at least ten courses of shingles above the tie-in area you can start laying out for the tie-in.

First, notice how the five inch pattern bond lines fall into rows at a slight angle. Select one row that will come down to the eave near the bottom of the valley, as in Figure 5-30. This row does not have to be at the end of a shingle. At the top you need to drive a nail so that you can anchor the end of the chalk line. If this is not at the end of a shingle, lay a shingle on top of this row and use this shingle as a guide. At the bottom of these shingles you will always need to lay out a shingle on top of those already placed to determine the top of the shingle directly above the bond line. Note Figure 5-30. Now hook the chalk line at the top corner of the top shingle and bring the line down to the top corner of the lower shingle. Take this length of chalk line and remove the chalk by snapping the line several times on a portion of the roof that will be covered later. Now stretch the line from the top shingle, past the top corner of the bottom shingle and on down to the bottom edge of the roof. Snap the chalk line leaving chalk on the felt paper only. In Figure 5-30 the bottom edges are even across the roof so the measurements would be the same on both sides of the dormer. Therefore, you could take the chalk line back to the left side of the dormer and stretch it out like you did on the tie-in area. The bottom shingle on the left side would fall in the same position relative to line B as the bottom shingle on the right side would fall in relation to line A. Actually it is much easier to simply measure each time to determine the location of the lowest course of the tie-in side. In this case the measurement was 55 inches, an odd number, which means you must start out with the first shingle *not* on the line. With this method the chalk line will line up with a definite part of the top of each shingle every six inches starting at either end, even though the shingles are on a five inch pattern.

*Shingle Tie-ins*

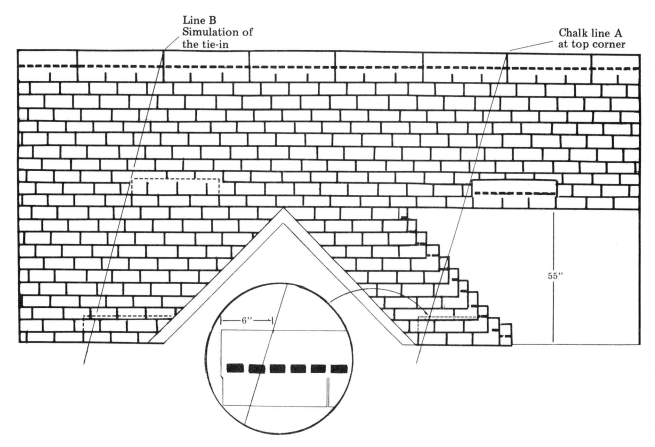

5 inch pattern tie-in
Figure 5-30

There are only two positions where you can begin. One is directly above a bond line, which is the same as the top corners. You can also start out with the chalk line at the top of the shingle half way between the bond lines. There is usually a factory notch cut into the shingles at this point. One notch is six inches in from each end and the third notch is exactly in the middle of the shingle. These notches are only slight cuts at the top of each shingle. You may have to turn the shingle over to see it. To make this cut easier to work with you can bend one side of the cut one way and the other side the opposite way. If the measurement gives an odd number, start the first shingle with one of the notches on the chalk line. See the inset in Figure 5-30. If the number is even, such as 50 or 60, start with one corner of the shingle on the chalk line.

After you find which shingle to start out with you can install the starter course and commence shingling again. As you shingle back up you should use the chalk line as often as possible to keep the bond lines in line with the other shingles. Sometimes as you shingle back up the roof the bond lines will not line up perfectly straight when you gauge the shingles for the five inch offset. When you lay a shingle on the chalk line you should position the shingle with a notch or top corner on the line. Only when the shingle hits the chalk line directly above one of the two bond lines in the center of the shingle do you need to gauge the five inch offset, as in Figure 5-31. This is the second shingle laid in the tie-in. The sixth shingle will be the last shingle to hit the chalk line on this row of shingles. You can either gauge the rest of the way or you can go back down to the bottom and bring up another row of shingles.

Had the bottom edges been uneven you would measure straight down as was done in Figure 5-4 and place marks at five inch intervals down to near twelve inches from the bottom. Again, if the marks give an even number of courses, place a shingle with the top corner on the line at the lowest mark and fill in under it accordingly. If the number of courses is odd, start out with a notch of the first shingle on the line.

For a small tie-in with a five inch pattern where you have only five or six shingle courses involved, back down from the other shingles as

Roofers Handbook

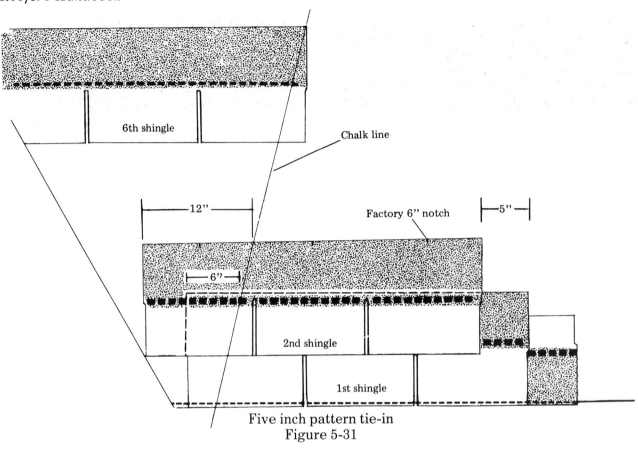

Five inch pattern tie-in
Figure 5-31

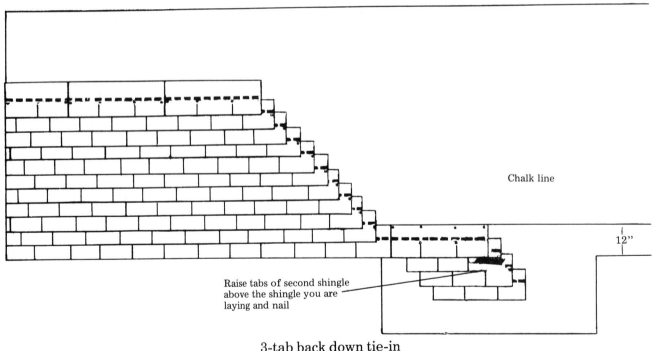

3-tab back down tie-in
Figure 5-32

in Figure 5-32. This method is fairly fast for a short tie-in. On a long tie-in you can get the shingles crooked very easily. To back down, chalk a straight line out over the tie-in area so that you can lay one more row of shingles. From the bottom shingle, which should be nailed at the top only, slide one shingle up underneath and gauge each end for the five inch exposure. Then gauge the five inch offset. Now you can raise the tabs of the second shingle up from this shingle and place three nails one inch up from the top of the bond lines. These nails should penetrate the top of the shingle that you just laid. Each time you lay a shingle you should go up two shingles and place three nails. After you reach the bottom you will need to install the starter and fill in the required shingles on the left side of the shingles that you just backed down. Then you can complete the proper nailing of the shingles. Always be careful that you don't leave out any nails. Now continue shingling the slope as the tie-in should be finished.

# Chapter 6
# Flashing Fireplaces and Chimneys

The point where the roof surface joins the fireplace or chimney is one of the most common areas for leaks to develop. Fortunately, stopping leaks here is not difficult. A few simple precautions will prevent most leaks. Always be careful about the application and the selection of the material to be used. Figure 6-1 shows the proper installation of a galvanized metal flashing, usually referred to as a saddle type flashing. A metal flashing like this can be custom made by most sheet metal shops from a drawing or description of the roof and chimney. All that is needed is the roof pitch and the width measurement. Proper installation is still important to ensure a watertight roof. The metal flashing must be positioned at the exact edge of the fireplace opening so the bricklayer can keep the bricks tight against the metal. The roofer must shingle under the bottom edge properly so that the water will run out onto the roof. The valley portion must be trimmed so that there is a water trough in the valley where the water can run smoothly, and the corners of the valley shingles must be dubbed. This is explained more fully in Chapter 7.

The sides of the fireplace should be flashed with galvanized metal as in Figure 6-2. This can

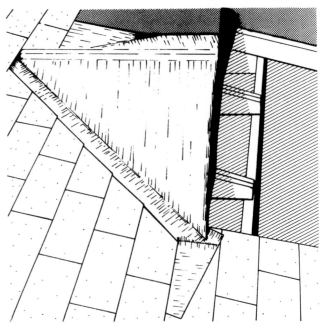

Proper installation of galvanized flashing
Figure 6-1

*Flashing Fireplaces and Chimneys*

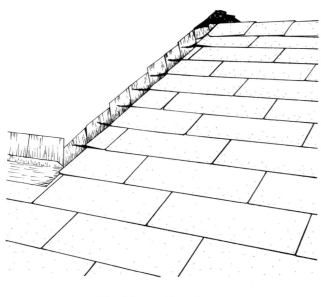

Flashing a fireplace
Figure 6-2

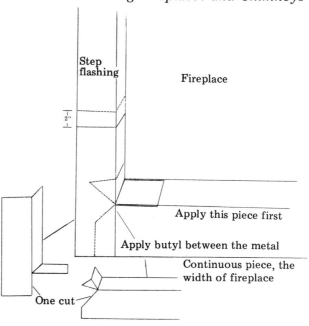

Fitting metal fireplace flashing
Figure 6-3

be installed as you shingle around the fireplace but it is easier to put the metal in after shingling is completed. Go ahead and shingle this small area, leaving out any nails in the shingles where the metal is to be installed. Chalk a line on the shingles about ¼ inch from the edge of the wood opening and then cut the shingles on the line. Now go back down to the bottom of the fireplace and start installing the metal flashing. The corners should be a soldered one piece unit, but usually the roofer has to make the corners watertight by making neat and exact cuts. See Figure 6-3. Always use a good grade of caulking such as butyl rubber sealant to permanently seal the unsoldered corners. Overlap each piece of metal at least two inches. This metal flashing should extend out onto the shingles at least four inches and turn up about four inches so that the counterflashing will be able to lap down over it with at least three inches.

Any time that the horizontal width of the chimney is over two feet you should install a galvanized metal saddle flashing as in Figure

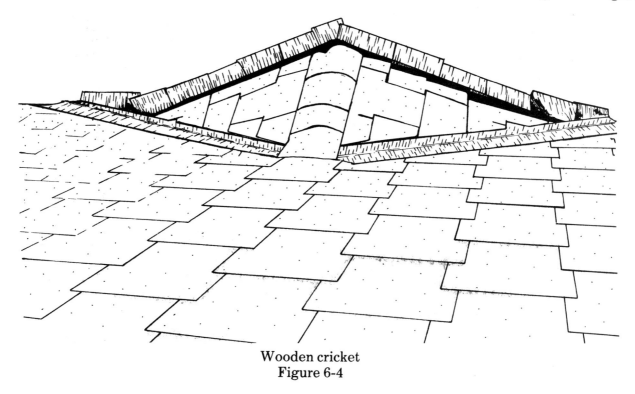

Wooden cricket
Figure 6-4

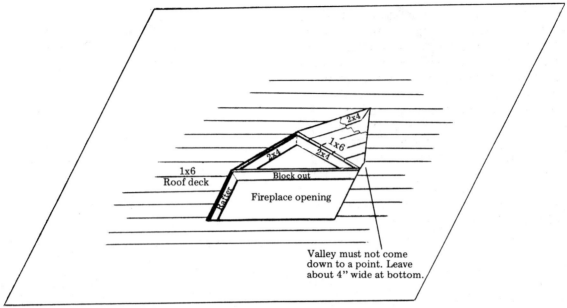

Wooden cricket structure
Figure 6-5

6-1. If a galvanized metal saddle flashing is not available the roofer or carpenter should build a wooden cricket as in Figure 6-4. This can be built very easily using scrap lumber. See Figure 6-5. To help eliminate future problems, the cricket should have the same pitch as the main roof. The bottom of the cricket, where the valley comes down against the chimney, should never be pointed. Instead, the bottom edge should be about four inches wide so that a water trough can be formed in the valley. If the cricket is built with the roof deck coming to a point at the bottom, there is no way that a water trough can be formed. The valley will be "pinched" at the bottom, creating a possible leak from leaves damming up and detouring the water. The best valley material to use on a chimney is the "W" shaped formed valley. Using this type of valley is especially important when you have a wide horizontal chimney at the bottom of the roof. Here the water will come off the roof in a large volume and with more force. The W shaped valley is designed to turn this water downward immediately. With a smooth valley the water will rush under the shingles on the cricket and pass through the roof. Always be careful when you shingle the cricket to see that no nails are driven into the metal valley more than two inches from the outside edge of the metal. When installing the metal valley, place one foot in the center at the bottom to hold it down in position while you nail the edges.

Not all fireplaces and chimneys present the same problems to the roofer. However, the three chimneys illustrated cover most situations. Just remember that neatness and thoroughness are the keys to making a watertight chimney joint.

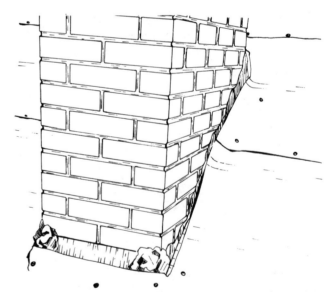

Chimney ready for flashing
Figure 6-6

### Flashing On Reroofs

When reroofing, it is almost always better to use new metal flashing and install it up under the existing metal counterflashing whenever possible. However, in many cases you won't be able to do so and you must do the best that you can under the circumstances. Most homeowners

## Flashing Fireplaces and Chimneys

and contractors do not want to go to the extra expense and trouble to saw out the mortar joints of the chimney with a carborundum blade and install counterflashing. Consequently, you must seal the chimney or fireplace another way. Figure 6-6 shows a tear off job with plywood nailed on over the old roof deck and asphalt felt already nailed on. There was no counterflashing to tie in under so the roofer had to install new metal flashing and then seal the top with plastic cement. See Figure 6-7. Then the roofer laid a four inch strip of membrane in this plastic cement and smoothed it out. The membrane was coated with a 1/8 inch layer of plastic cement and then spray painted with a good grade of aluminum paint. When you apply the first layer of plastic cement, enough of it will squeeze down in and around the metal flashing to bond the chimney and metal together. The membrane and top coating of plastic cement will help keep this bond from cracking later when fully cured. The aluminum paint will simply prolong the life of the entire flashing.

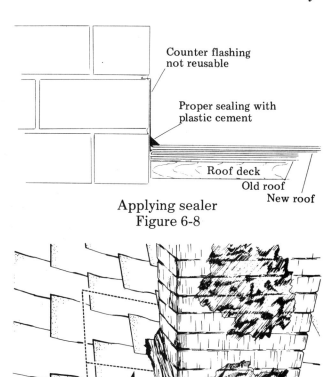

Applying sealer
Figure 6-8

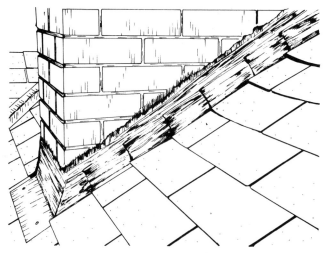

Chimney with new flashing and sealer
Figure 6-7

If you are reroofing and you have a smooth surface to tie in to, you may be able to seal off the shingles with plastic cement. To make this last as many years as possible, carefully apply the plastic cement about 1½ inch wide at a 45 degree angle. See Figure 6-8. Leaving a ¼ inch gap between the shingles and the chimney will permit more plastic cement to seal the edge of the shingle. The best tool for applying the plastic cement is a wood shingle. Break about four inches off of the tail end of a shingle and round the corners slightly by rubbing it on concrete. Many roofers refer to plastic cement as ''bull'', and the wood shingle trowel as the

Flashing on a new roof of an old house
Figure 6-9

''bull paddle''. This method of sealing requires that the plastic cement be painted with aluminum to give protection against the elements.

Figure 6-9 shows a typical reroof on an old house. The chimney and the counterflashing are old and rough. To prevent the roof from leaking the roofer had to apply a wide band of plastic cement and membrane. This is never very pretty but in this type of situation there is no other choice. At the top of the fireplace or chimney at arrow A there is always one shingle or tab that should be raised and sealed under to prevent leaking. Water will run under this shingle or tab at an angle. When the water gets down to the plastic cement it will run under the shingle and leak. When you seal under this tab or shingle you should seal the joint in the shingles below as high as possible. See arrow B, Figure 6-9. At arrow C is another possible leak in a hard blowing rain. Sometimes this edge is closer to the chimney and should definitely be sealed off by raising the tab and applying plastic cement. Whenever you raise a tab or shingle and seal underneath, press the shingle down firmly and then seal on top also.

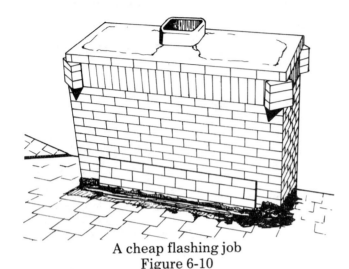

A cheap flashing job
Figure 6-10

New flashing sealed with plastic cement, membrane and aluminum paint
Figure 6-11

## Stopping Leaks On Chimneys

Figure 6-10 shows a classic example of a cheap "hurry up" job. No metal flashing was installed around the chimney when the house was built. The roofer made several attempts to repair the leaks but did not succeed. The leak problem was helped somewhat when the shingles were removed around the three sides and metal flashing was installed. The metal extended under the shingles twelve inches and eight inches up the side of the chimney, where the top edge was sealed with plastic cement, membrane and aluminum paint. See Figure

Brushing on a coat of silicone sealer
Figure 6-12

6-11. The fireplace still leaked but somewhat less. Needless to say, the homeowner was more than slightly irritated. The next step was to prevent any possible leak at all. A flu cap was installed at the top of the chimney to prevent water from running down inside. Sometimes water will run down the flu tile until it comes to an open joint, then detour into the sheathing and appear as a roof leak. At the top of this chimney the mortar was cracked, permitting water to enter the interior of the fireplace itself. This could also cause what might appear to be another roof leak. The mortar was sealed with plastic cement, membrane and aluminum paint. To prevent water from soaking into the mortar joints and getting behind the counterflashing, the sides of the fireplace received a brush coating of clear silicone sealer. See Figure 6-12. This type of sealer can be purchased at most lumber yards.

The first attempts at repairing the leaks around this chimney were poor efforts to get by at minimum cost. Applying plastic cement all over the shingles seldom works and certainly makes it harder to repair the leaks properly later. This chimney should have had metal saddle flashing installed in the first place. The builder of the addition did not realize this and the roofer should have counseled him about the flashing needed. As a professional roofer you will be aware of potential problem areas and poor construction techniques that can cause serious headaches in the future. You have an obligation, as a professional, to point out what is wrong to the contractor or owner. Even on the most minimum "no frills" job, there are few builders or owners who won't spend a few dollars more to prevent leaks during the first rainstorm. If you ignore this obligation your own reputation as a craftsman may suffer when the roof fails.

# Chapter 7
# Valleys

The potentially most troublesome part of a roof is the valley. This is where the greatest pressure from water or snow will build up and create the most favorable conditions for a leak. Not just any valley will work on every roof. Every roof has its own special needs. There are basically four different types of valleys; the full lace, the half lace, the W shaped formed valley and the smooth valley which uses either metal flashing or a double layer of 90 pound roll roofing the same color as the roof. When selecting the right type of valley, examine the situation carefully and think about how the water will run. If the valley has a definite curve, obviously a smooth valley would be wrong because the water would run under the shingles at the turn. Here a half lace valley would be right. Sometimes on a large roof there may be a need for more than one type of valley.

### The Full Lace Valley

This is one of the oldest types of valleys and in certain parts of the country it is gaining popularity again. See Figure 7-1. The reason for the current popularity is that this valley requires no trimming. The shingles are laced back and forth up the slope. Normally, when both slopes are the same pitch, there will be an equal number of shingles on each slope. Then it is a simple matter to slip one shingle under the adjacent shingle on the opposite slope. In Figure 7-1 the valley is a junction of two different pitches. One slope contains fifteen shingles and the other slope, which is flatter, contains thirty six shingles. To work the shingle courses together, two shingles of the flatter slope were slipped under the same shingle on the steeper slope. Occasionally a third shingle was slipped in as the roofer worked up the valley.

When you shingle up the valley on the first slope you must leave the shingles loose in the valley so that you can slip shingles in when you shingle the other slope. The shingles in the valley should always be whole shingles. When a shingle ends in the valley, slide a whole shingle over toward the valley and nail on a single tab or maybe a two tab shingle in the gap between whole shingles. Having whole shingles in the valley area is important because there should be no joints in the valley. Water will run under the tabs and leak through a joint. If there is any doubt about a joint being too close to the valley, raise the tab and seal the joint with plastic cement.

The main objection to a full lace valley is that the appearance lacks the professional look. There is nothing neat about it and many homeowners would prefer not to have this type of valley on their house. Also, in a full lace valley it is very difficult to find and repair a leak.

Full lace valley
Figure 7-1

Full lace valley on a lock shingle roof
Figure 7-2

Smooth metal valley with half lace valley
Figure 7-3

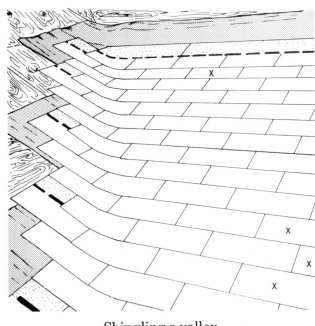

Shingling a valley
Figure 7-4

To guard against leaks you should apply a layer of metal or heavy roofing paper on top of the 15 pound cap sheet before applying the shingles. The full lace valley can be used on most roofs that have a pitch of 3 in 12 or steeper. Note, however, that the full lace valley should never be used on a lock shingle roof as in Figure 7-2. There is no way to waterproof a valley like this.

### Half Lace Valley

Figure 7-3 shows a half lace valley at the bottom with a smooth metal valley at the top. The smooth valley was used in four other places on this roof, but at the bottom of the curved valley a half lace valley was installed so that there would be no leaks. Water is very difficult to turn and with a smooth valley the water would run under the shingles on the flatter slope and result in a leak. With a half lace valley there is no open edge for the water to enter. Therefore, no chance exists for a leak. In Figure 7-4 you can see the first half of the valley. The shingles on the flatter slope are turned up on the other slope about twelve inches, more or less. When placing shingles in the valley, be sure that you keep the joints and nails at least ten inches from

# Valleys

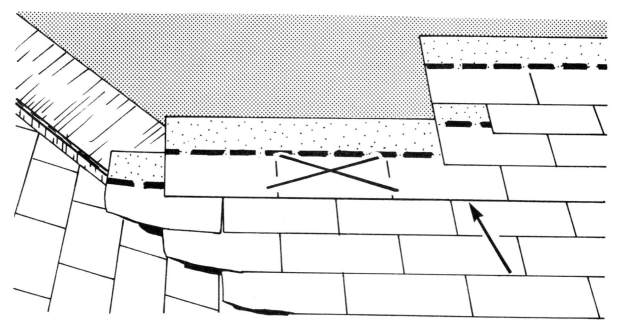

Shingling a metal valley
Figure 7-5

the center of the valley. In Figure 7-4 the shingles marked with an X are single tabs that were installed so that the whole shingle would not end in the valley.

Figure 7-5 shows a smooth valley being shingled. The shingle with an X on it will have to be moved over toward the valley and a single tab installed where the arrow is pointing. This precaution takes no extra time as the single tab is needed anyway. All that you will be doing is placing the joint in a better area away from the valley.

When you finish shingling the other slope you can trim the shingles back from the center of the valley about 2½ inches. This way the water will not run against the edges of the shingles.

Before you start shingling a slope, look the roof over and determine exactly which way the half lace valley should be applied. In Figure 7-3 it is obvious that the water will be running down the steeper slope on the left out onto the flatter slope. If the half lace valley were used in reverse, the water would rush down against the open edge and cause a leak. The half lace valley is the most universal of all the valleys. It can be used on any roof using a 3-tab or a strip shingle. It should never be used on a lock shingle roof. A layer of heavy roofing paper or a strip of flat metal can be used under it but is not necessary if the valley is applied correctly.

### W Shaped Formed Metal Valley

Figures 7-6 and 7-7 show a W shaped formed metal valley with a water guard at the edges. Normally, on a reroof, you should not use a valley with a water guard because the edges will not lay flat enough to make a smooth roof. On this particular roof it was used because formed metal flashing without the water guard was not available. When using the metal valley with a water guard, which is used on most commercial roofs, you always use metal clips about every twelve inches to hold the metal in place. In Figure 7-6 the metal itself was nailed to hold it tight against the shingles.

The W shaped metal gets its name from the one-inch hump in the middle of the valley. You can use this valley on almost any roof except where you have a sharp turn in the valley. It is hard to turn a corner with this type of flashing because of the hump in the middle. If it is necessary to use this metal in a curved valley you can make a cut on each side of the hump at the curve and then solder the metal together again after it is bent. When you shingle into the valley, always be sure that you don't nail more than two inches past the edge of the metal. If a tab is needed in the valley you can use the method illustrated in Figure 7-5. To join together two pieces of valley with a water guard at the edge, as in Figure 7-6, you will need to bend the edges of the bottom piece back and then slip in the top piece. After aligning the metal and nailing it you can bend the metal back into place. In the center of the valley at the top in Figure 7-7 is a sharp hump that should be bent down with a hammer before the ridge is applied. When you trim the shingles, leave

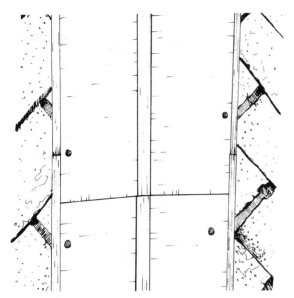

W shaped formed metal valley
Figure 7-6

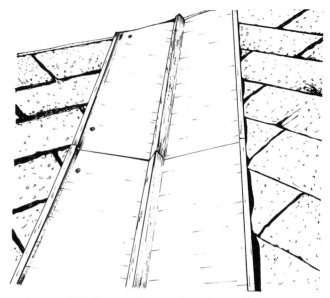

W shaped formed metal valley
Figure 7-7

about 2½ or 3 inches on each side of the hump for the water trough.

### The Smooth Valley

A smooth valley, using either a double layer of 90 pound roll roofing or a strip of galvanized metal is probably the most widely used valley. The metal should be twenty inches wide so that you have ten inches of metal on each side of the valley. To apply the double layer of 90 pound roll roofing you should first measure the length of the valley and then cut that length plus about one foot from the roll. Now lay this out on the roof somewhere and cut it into widths one foot wide and two feet wide. Take the narrow strip and lay it in the valley with the granule side down as in Figure 7-8. After lining it up in the valley, nail one edge about every foot. Now force the strip down into the valley and nail the other side in the same manner. Using an old glove or a trowel, apply a line of plastic cement to this strip about nine inches wide and at least 1/8 inch thick. Lay the 2 foot wide strip with the granule side up as in Figure 7-9. Be sure to force this strip down into the plastic cement before nailing. Without the plastic cement this valley will not last as long as the roof. The sun will shrink the top layer of roll roofing resulting in the top layer being suspended without any support under it. It will then both wear out much faster and puncture easily from foot traffic or hail.

A smooth valley cannot be used on every roof. It should not be used in a curved valley or where you will have a larger amount of water entering the valley from one of the two slopes.

Also it should not be used if the structure of the valley where the roof decks join does not form a straight V shape the full length of the valley. If there is an irregularity in the slope of the roof surface there will be a hump in the valley that will probably detour the water under the shingles and cause a leak. The double layer of 90 pound roofing, when applied correctly, is approved by most building codes and will make an excellent valley. Regardless of the color of the shingles you can usually get 90 pound roll roofing in the same color. This roofing comes in a roll three feet wide, thirty four feet long and weighs 90 pounds. When you trim the shingles

Smooth 90 pound valley
Figure 7-8

*Valleys*

Laying a smooth valley
Figure 7-9

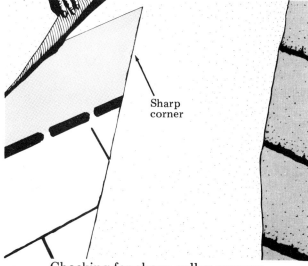

Checking for sharp valley corners
Figure 7-10

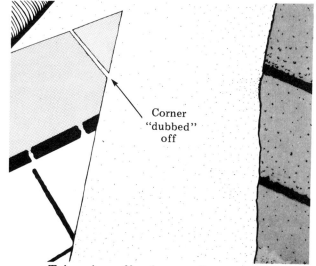

Trimming off a sharp valley corner
Figure 7-11

you should use a hook blade and be careful not to cut into the valley material. Leave about a five inch water trough at the top and bottom of the valley.

## Valley Tips

After trimming the shingles in the valley, go back up the valley raising the shingles one by one and trimming off the sharp valley corners. Note Figures 7-10 and 7-11. Failure to trim these corners is one of the biggest reasons for roof leaks. Any water that runs against an edge of a shingle will hit a sharp corner and run along the top of the shingle until it finds a place to leak

A leaky valley
Figure 7-12

through the roof. Sometimes the water will run for many feet before it finds an opening in the felt underlay. Trimming off these corners is referred to as "dubbing" the corners.

Many roofers don't understand the importance of trimming these corners because they work for more than one company and are never called back on their leaks. Many roofing companies fail to keep performance records and have no way of knowing which roofers are leaving valley corners untrimmed. Sometimes it may take two years or more for the shingles to lay down tight enough to leak. This depends on the type of valley and whether there are any dust accumulations or granules under the shingles. Once the water gets under the shingles it can cause a leak almost anywhere. On one particular roof the water ran half way down the valley, hit one of these sharp corners and traveled about twenty feet, leaking along the way through two different rooms. Then, if the rain continued long enough, it would run a few more feet to a hip, run down the hip until it hit the top of another shingle on the other slope, and run until it leaked into the third room. This appeared to be three separate leaks though it was caused by one defective valley.

In Figure 7-12 a tear off job is in progress. You can see where the roofer had laid a 3-tab shingle in the valley lengthwise (instead of trimming the corners) in an effort to prevent the valley from leaking. This was a half lace valley so it only leaked on one side. However, as the dirty water trail shows, it leaked from one end to the other.

When you trim out and "dub" the corners you are actually building a trough for the water to flow down to the eave. Therefore you must center this trough so that the water will run in the center of the valley. When reroofing on a composition shingle roof there will be enough loose granules in the valley to help you to determine the exact center. Note Figure 7-13. You can either raise the shingles and look or you can cut back the excess shingle until you reach the center and then mark where you want to cut. A nice width for a valley is about five inches at both the top and bottom. Sometimes, after you have cut both sides, you can tell by the granules that the valley is not straight. You may have to cut one side again to keep the water from running under the shingles.

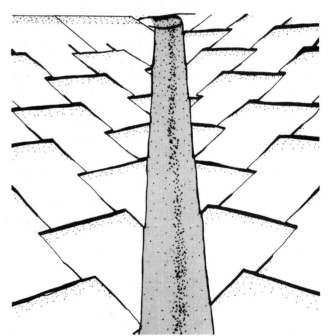

Loose granules in center of valley
Figure 7-13

If you have no one to help you chalk line the valley, instead of driving a nail to anchor one end of the line you can make a short cut in a shingle edge at one end and drop the knotted line end in it. Make the cut at the top of the shingle in the same line that you will be cutting. In Figure 7-14 you can see how easy it is. On the opposite side, after making a cut and stretching the chalk line out, it was decided that the cut should be in a little closer to the center of the valley. The ridge covered this cut later.

When you are shingling into the valley, especially with T-lock shingles, be careful that

# Valleys

Chalking a valley
Figure 7-14

you don't nail on shingles that are not needed. In Figure 7-15 at the X it looks as if another shingle is needed. Simply place a shingle in the next position, as in Figure 7-16, to determine if one really is needed. As you can see, no shingle is needed. If the unneeded shingle had been applied the roofer would have placed two nails in an area that should not be nailed through.

If you suspect a problem in a valley, perhaps from a large amount of snow drifting onto the roof or because the valley is crooked, you can guard against leaks by applying a liberal amount of plastic cement to the edges of the valley material. See Figures 7-15 and 7-16. This will form a barrier when the shingles are applied. When nailing the shingles in this area, be sure not to nail past the plastic cement. When you nail through the plastic cement, do not overdrive the nail, just bring the nail down snug. The corners will still need to be "dubbed".

When you are shingling into a valley, try to keep all of the nails back as far as you can. Try to nail within the first couple of inches of the valley material, as the line in Figure 7-17 indicates. When using a W shaped formed valley with a water guard, keep all nails out of the valley. To secure the metal in place, use metal clips. This nailing procedure will help eliminate leaks and make it easier to lift the shingles and "dub" the corners.

### Waterproofing A Blind Valley

Figure 7-18 shows two valleys coming down to one flat valley (blind valley). This is always a problem because of the large amount of water that converges on the blind valley. In the winter an excessive amount of snow may build up between the two valleys. The old roof has just been removed and now the roofer's task is to install a watertight roof.

The first step is to select the right type of valley. Here a double layer of 90 pound roll roofing was chosen because it could be laid in one continuous strip, thus eliminating a joint that would have been necessary with metal. The valley here is over ten feet long and valley metal usually comes only in ten feet lengths. Also, metal will expand and contract with temperature changes, breaking any bond or seal with the shingles and the blind valley. The blind valley

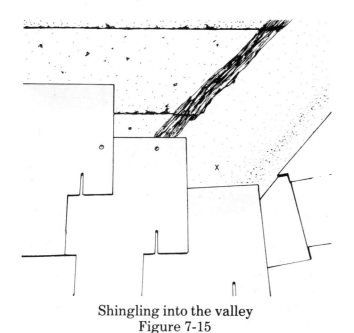

Shingling into the valley
Figure 7-15

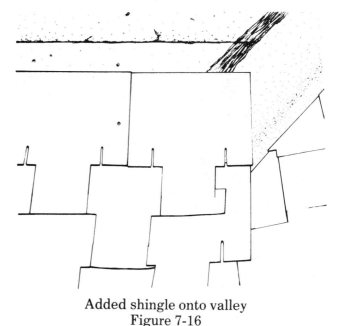

Added shingle onto valley
Figure 7-16

here had a low spot in the center. So, to start out, a two foot piece was cut off of the roll and laid in the valley. Then a two foot six inch piece was cut and laid in the valley, overlapping onto the other piece about one foot. See Figure 7-19. The lap was sealed with plastic cement. This provided an extra layer in the low area. Plastic cement was then applied to these two pieces that were laid upside down. Now a six foot strip of 90 pound roofing was laid in the blind valley, making sure that it was rubbed down well into the plastic cement.

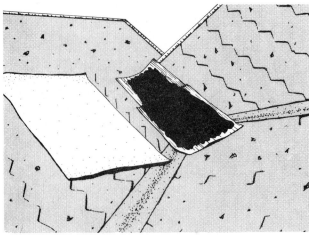

Sealing the blind valley
Figure 7-19

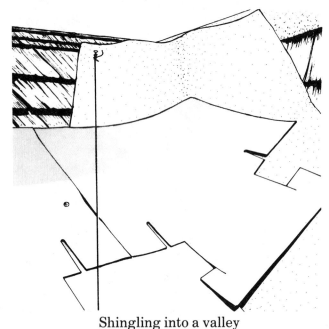

Shingling into a valley
Figure 7-17

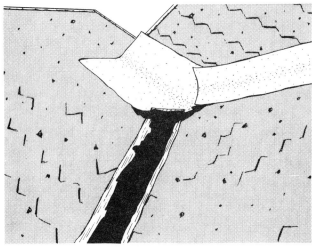

Laying the valleys
Figure 7-20

Two valleys coming to one flat valley
Figure 7-18

In Figure 7-20 the valleys are being laid; first, the one foot strip upside down with plastic cement troweled on and then the two foot strip, granule side up, over it. Where the one foot strip and the two foot strip lapped onto the blind valley, plastic cement was applied to seal the two valleys together and prevent water from backing up under the roofing.

Before the roofer started shingling, an extra precaution was taken since the homeowner always had a leak in this area. Plastic cement was liberally applied to the edges of both valleys and three sides of the blind valley, as in Figure 7-15. This will prevent any normal freeze back that occurs in the winter. After the shingles were trimmed in the valleys, the corners were dubbed. This double layer of 90 pound roofing made an excellent valley, but be careful not to step in it. A sharp heal will puncture the roll roofing until the cement hardens. After the plastic cement cures it will become much stronger.

Improper Valley Material

Selection of the proper valley material is always important. Never use a smooth valley if

*Valleys*

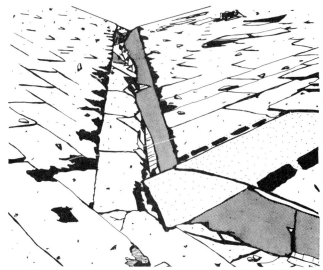

Crooked wood structure in valley
Figure 7-21

Rust in the valley
Figure 7-22

Wrong choice of valley material
Figure 7-23

Right choice of valley material
Figure 7-24

the wood structure of the valley is crooked, as in Figure 7-21. The 90 pound roofing will buckle and detour the water under the shingles. The rusty water trail in Figure 7-22 is clear evidence of a leak due to a crooked valley. A half lace valley should have been used so that when the water ran down to the buckle it would simply run out onto the roof instead of under the shingles.

When you use a half lace valley, always be sure that you shingle the correct slope first. Shingle first the slope that will receive most of the problem water.

Figure 7-23 shows a situation where the wrong valley material was used. This is a new roof with a smooth valley. You can see that there is one steep slope which will carry more water and the water will be running much faster into the valley. This water will rush against the water from the flatter slope and force it back under the shingles. This resulted in a very large leak which the roofer tried to stop with plastic cement. This having failed, the shingles were removed on both sides of the valley and a W shaped formed valley was installed. The ridge in the middle will keep the water from the steep slope from rushing across the valley and under the shingles. This is the only type of valley that could have been used on this roof. If the shingles had been 3-tabs or strip shingles, a half

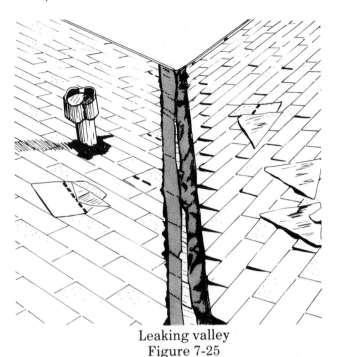

Leaking valley
Figure 7-25

Sliding the starter piece of metal
valley into place
Figure 7-26

lace valley could have been installed, shingling the slope on the left first. Any time that you have crooked valley construction, a deliberate curve or a steep slope coming down to a flatter slope, use a half lace or a W shaped formed valley.

### Replacing A Valley With A W Shaped Formed Valley

Figure 7-25 illustrates a leaking valley where an addition was built on to the main house. The old valley was covered with plastic cement so it was removed by pulling the nails on both sides and sliding the metal down and out. When the nails were removed a few shingles were dislodged. These were laid on the roof out of the way. Since the valley was 15 feet long, a six foot starter piece was used at the bottom. You may need to position a ladder at the bottom of the valley to get the metal started under the shingles. Once you get it started you can slide the metal up into the approximate position. Figure 7-26. Now you can start the whole piece of metal about one foot down from where the short piece left off, as in Figure 7-27, and slide it up into position. Once you get the two pieces of metal in position you can line them up in the center of the valley and nail them in place. Finally, replace shingles as required.

The finished valley in Figure 7-28 is now watertight. Also note that there is a vent near the valley. Any time that you are repairing a leak in a valley, check any vents or chimneys that are in the area. Sometimes a vent or

Sliding the rest of the valley into place
Figure 7-27

chimney leak will show up at a valley and a valley leak will appear someplace else. Take nothing for granted when you are looking for a leak. In this particular situation both the vent and the valley were leaking. The valley leaked because the corners were not trimmed and the vent leaked because the flashing was installed incorrectly.

### Replacing A Valley With 90 Pound Roll Roofing

Sometimes a 90 pound roll roofing valley will

*Valleys*

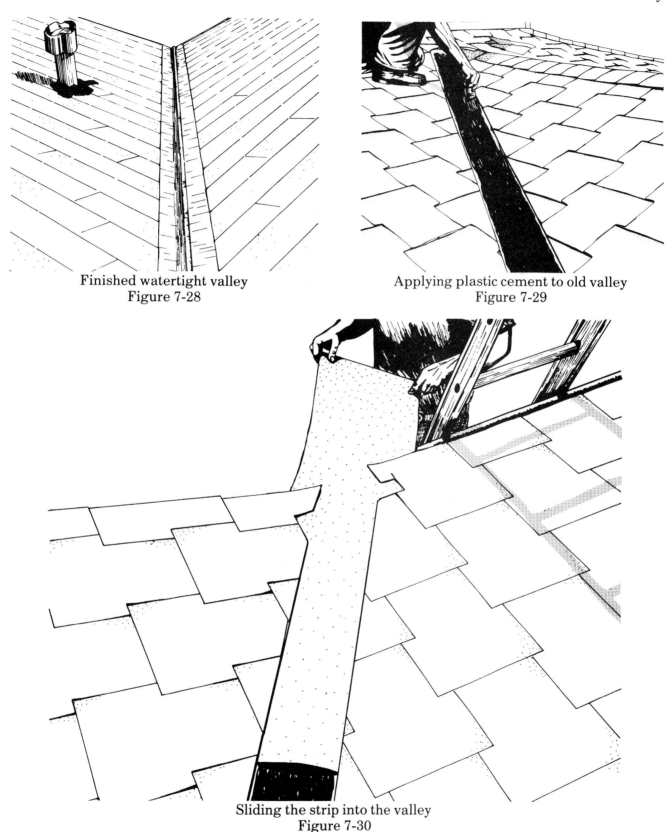

Finished watertight valley
Figure 7-28

Applying plastic cement to old valley
Figure 7-29

Sliding the strip into the valley
Figure 7-30

be destroyed by hail because it was installed incorrectly. In this case it is better to leave the old valley in and just add to it. First, remove the nails on both sides of the valley to about 12 to 14 inches from the center. Any loose shingles should be laid up out of your way. Now cut an 18 inch strip of 90 pound roofing the exact length that is needed. The next step is to lift the shingles on each side of the valley and make sure that you have all the nails removed. Apply

Chalking a line for a new valley
Figure 7-31

Shingles cut and ready for the valley
Figure 7-32

plastic cement, about ¼ inch thick, to the old valley. Get as close to the edge of the shingles as you can. See Figure 7-29. Slide the 18 inch wide strip under the shingles, starting at the very bottom. It will probably take two men to install this valley, one to pull the strip up into position and the other to hold and feed the material to him, as the roofer is doing in Figure 7-30. After the material is in position, you can renail the shingles and replace any shingles that were removed. Be careful that you replace all the nails that were removed. An open nail hole will result in a leak. Make sure that all corners have been dubbed. If the shingles are too hot to cut easily, remove the corners by tearing them off by hand.

The procedure outlined here can also be used to cover up a rusty metal valley.

### Cutting In A Valley On An Addition

Many valley problems begin when an addition is built on to a house and a new valley must be cut in the roof. The best way to handle this is as follows: Measure out from the wood sheathing three inches at the top and bottom and chalk a line as in Figure 7-31. This particular valley had to have a second line about one inch further out because the wood sheathing was a little crooked. Now make two cuts on each shingle, the first cut on the chalk line and the second cut at an angle about one inch in from the line, as in Figure 7-32. This will make it easier to slip the valley into place. After cutting the shingles the length of the valley, start at the top and remove all the nails that are within 12 or 14 inches of the valley. No matter what type of valley you are using you should start at the bottom and slip the valley into place. This way you will be going with the laps and will not have to lift every shingle. After the new valley is in you can align it properly and, while forcing it down in the middle, nail it in place. Next, replace shingles as necessary. If you made the cut in the right place you won't need to recut the shingles. Make sure that no old nail holes are left open, and trim the corners of the shingles.

# Chapter 8

# Ridges

A straight and neat ridge will improve the appearance of any shingle roof. Fortunately, applying a good, professional ridge is relatively easy. First trim the shingles back out of the way so that there is no hump in the ridge. See Figure 8-1. At the very bottom of the hip place a tab and nail it in place. On a new roof use nails 1¼ inch long. On a reroof job the nails should be 1½ inch long. Now trim off the corners of the tabs at the bottom and place a 3 penny galvanized nail, as in Figure 8-1, so that you can anchor the end of the chalk line. Stretch the line on up to the top of the hip. You will use this line as a guide for aligning the ridge shingles. Using another ridge tab, estimate where you want the chalk line. Be careful when you snap the line that the wind does not blow the line to one side. Always raise the line straight up but not too high. Also, avoid using red chalk as red does not wash off as easily as other colors.

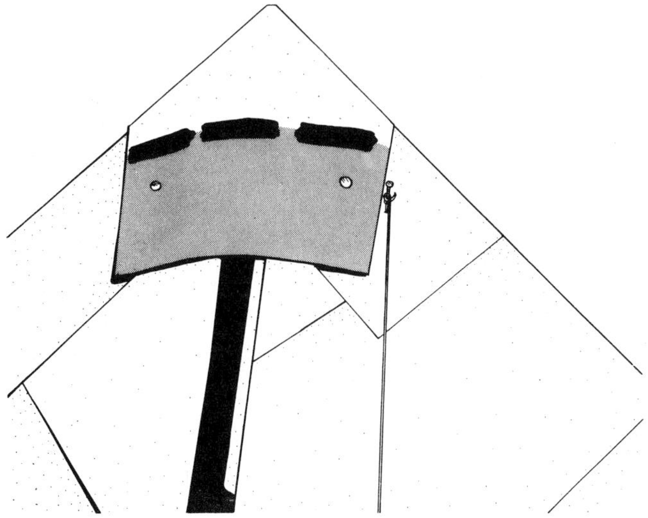

After snapping the line you are ready to start ridging. Start at the bottom and sit on the same side of the ridge as the chalk line so that you can align the tabs easier. The ridge will look much smoother if you nail about two inches above the factory adhesive. Note Figure 8-2. If you pre-bend the ridge singles, as in Figure 8-3, they will be much easier to lay straight. If the temperature is too cold, prebending the tabs may not be possible because they will break in the middle. In cold weather you should lay the tab on the line, nail that side and then bend the tab over the ridge. If they still break you may have to wait until warmer weather to apply the ridge. Do not try to stretch the ridge shingle exposure out too far. The ridge should have the same five inch exposure as a 3-tab roof.

Chalking the ridge
Figure 8-1

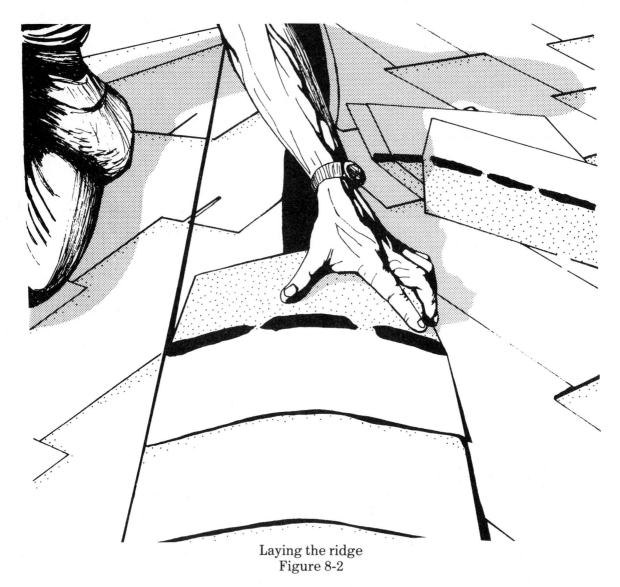

Laying the ridge
Figure 8-2

### Cutting The Hip Ridge Out Of 3-Tab Shingles

If you have to cut ridge shingles from 3-tab shingles you will probably need about four bundles for the average hip roof. This will be a lot of cutting so you should simplify and systematize the job as much as you can. Four bundles is about 107 shingles and each shingle must be cut in six different places. Therefore, 642 cuts are needed just for four bundles of shingles. To start out you need a work bench or an area that is a comfortable height for work. The tail gate of a pick-up truck will serve nicely. Tear the wrapper off the first bundle of shingles and pick it up with the tabs away from you and the granule side up. If you are right handed drop the left end of the bundle from about one foot. This will break the shingles loose on that end. Now hold your left hand on top of the bundle at the left end and let the other end fall. This will finish breaking the shingles loose and will also stagger the shingles so they will be easy to cut. See Figure 8-4.

Now cut off a slight amount at an angle, as in

Pre-bending the shingles
Figure 8-3

## Ridges

about four or five minutes, providing that you use a sharp hook blade.

Cutting off this excess is necessary if you are to install a neat ridge on a hip roof. For the ridge at the top of a gable roof it will not be necessary to cut the shingles in this manner. Instead, make two straight cuts on each shingle, directly below the cut-outs.

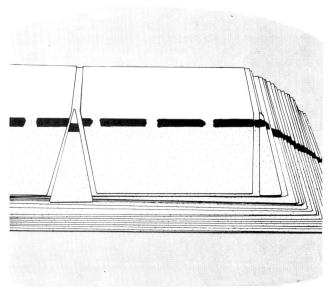

Cutting the shingles
Figure 8-4

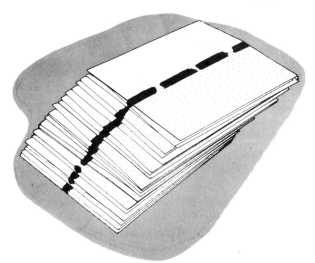

Leaning the shingles to the right for cut No. 6
Figure 8-6

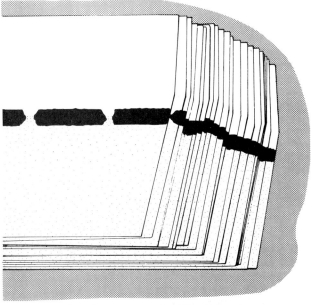

Cutting an angle (Cut No. 1)
Figure 8-5

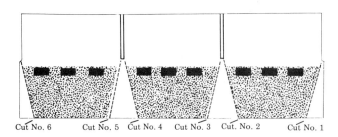

Cutting hip ridge
Figure 8-7

Figure 8-5. This will be cut number 1 in Figure 8-7. Hold down firmly on the tab that you are cutting. Now you are ready for cut number 2. As you cut each shingle, hold down on the top left corner of the tab. After each cut, simply toss the tab out of the way. As you toss each tab, try to make a halfway neat stack, but don't waste a lot of time doing so. Now you can make cut number 3 which will be the same as cut number 1. Then you can proceed to cut number 4. When you get to cut number 6 you will need to lean the shingles over to the right, as in Figure 8-6, before cutting. This whole procedure only takes

### Tips On Ridge Work

Any junction in the ridge should be clean cut and free of humps. Note Figure 8-8. Any face nailing should be done only at open cuts when needed and with a 4 penny or 5 penny galvanized nail. This is often referred to as "pin nailing".

Before ridging a T-lock roof, be sure that all the tabs have been installed wherever necessary at the top and on the hips. See Figure 8-9. This is necessary in order to complete the double coverage of the roof and also to cover the nails in the shingles. On a 3-tab roof you can sometimes use the scrap from a valley to fill in where it is needed on a hip.

Before ridging, make sure that excess shingle has been trimmed back from the hips and at the top. Note Figure 8-10. If a portion of a shingle is left on a hip it will cause a hump and leave one side of the ridge crooked. This is because when you lay one side on the chalk line, the shingles that are going over the hump will not reach as far as the other shingles.

Clean cut junction on the ridge
Figure 8-8

Excess shingle trimmed from hips and top
Figure 8-10

Tabs installed at the top
Figure 8-9

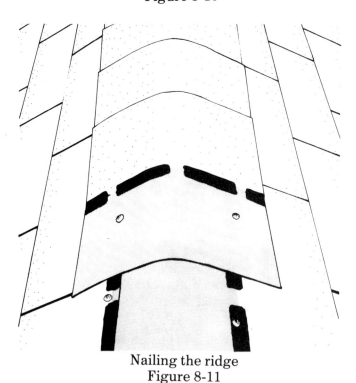

Nailing the ridge
Figure 8-11

Always nail the ridge on the upper portion of the shingle so that the next shingle will cover the nails. See Figure 8-11. This is called blind nailing. Face nailing, nailing in the exposed portion, will result in the nails rusting and working loose. There is no reason to face nail the ridge, especially when you are using self-sealing shingles.

When you are ridging a dormer, as in Figure 8-12, bring the ridge to the main slope so that the shingles will lap the right direction. Here the roofer had applied the ridge incorrectly and the last four shingles had to be removed and replaced correctly. Water rushing down the main slope will continue on top of the ridge far

*Ridges*

Repairing a leaky ridge
Figure 8-12

Removing an old ridge
Figure 8-13

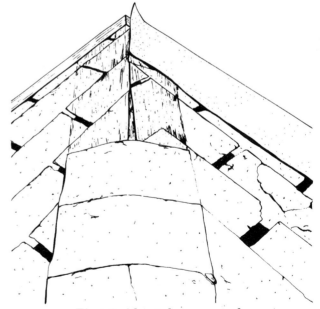

First 3 ridge tabs removed
Figure 8-14

enough to leak under the tabs if they are lapped improperly.

The old ridge should always be removed before reroofing to eliminate any unnecessary hump in the new roof. Figure 8-13. If the old ridge is left, the new shingles will turn up on both sides of the new ridge.

If you don't remove the old ridge, at least remove the first three or four ridge tabs at the bottom to prevent a hump in the new ridge. Figure 8-14. Usually the old starter is twelve inches wide. This results in a four layer build up in the old roof. Also, the ridge will have its triple coverage or head lap at the same area, thus

81

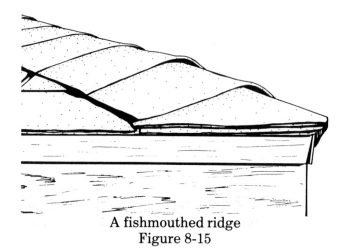

A fishmouthed ridge
Figure 8-15

causing a build-up of seven layers in one spot. When you install the new ridge, the second and third shingle will not lay down smooth. There will be a gap under the tabs, as in Figure 8-15. This gap is sometimes referred to as "fishmouth". Eliminate this build-up by removing the first three or four shingles at the bottom and trimming three inches off of the starter course.

Usually when you reroof over old wood shingles the ridge will be a metal ridge, as in Figure 8-16. If you remove the metal ridge, there will always be a few old wood shingles that will come off with it. You can go ahead and try to remove it. If you see that it will only make things worse you can leave it on. When you shingle up to it, cut the shingles off as close to the hump as possible. This will help cut down the height of the hump, thus allowing you to ridge over it smoothly.

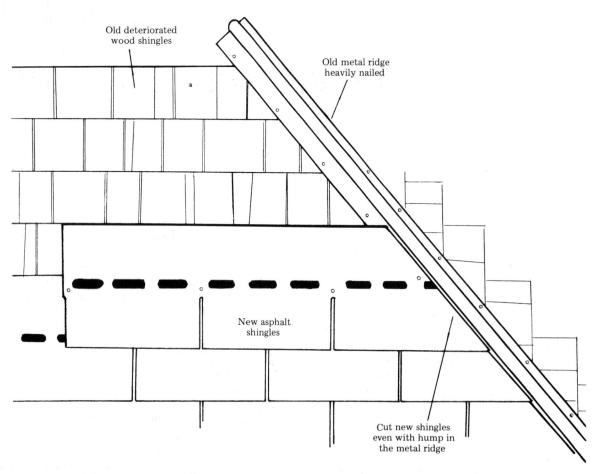

Reroofing around an old metal ridge
Figure 8-16

# Chapter 9

# Tips For Professional Roofers

Proper handling of materials will increase your production and make your job a lot easier. Three tab shingles are the easiest to lay out. Each bundle contains about 27 shingles and a bundle will cover about four strips of roofing felt. Begin by installing your starter course. Get into the habit of opening a full bundle of shingles and, without counting, grabbing the top one-quarter of the bundle with each hand. This should be about seven shingles. Toss these seven shingles on each lap of the felt as in Figure 9-1. The last fourth will probably be six shingles. Place these on the bottom lap. Laying out the shingles like this has several advantages. You should be able to reach a shingle at all times without moving very far and the shingles will be up ahead of you and out of your way. If you come to some shingles that are in your way, since they are on a lap you should be able to shove them up easily. If they are laid below the lap you will be shoving against the lap and will probably have to lift them up to move them. After a little practice you should be able to lay the shingles out as in Figure 9-1 with little effort. Keep the loose shingles close to the shingles that are nailed on so that you can reach them easily.

Many professional roofers adopt a position while shingling that prevents them from laying out their shingles to good advantage. You will

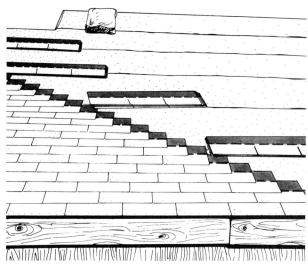

Laying out shingles in advance
Figure 9-1

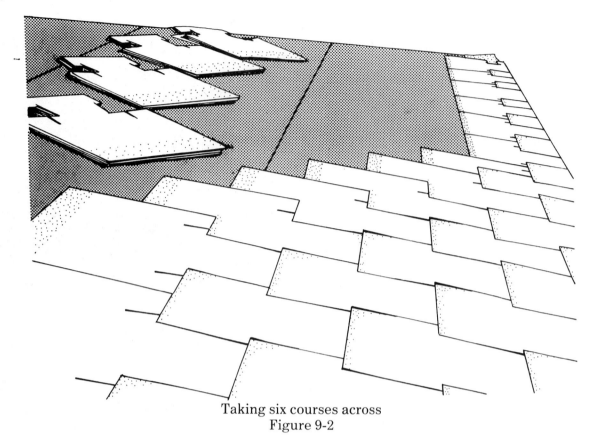

Taking six courses across
Figure 9-2

see roofers who work from the felt in a squatting position. Some bend over and others kneel down on their knees. The roofer that sits on his hip on the new shingles can lay out his work better and will nail on more shingles each day. Using the method described, shingles in each stack are in easy reach and the last shingle will be nailed in about the same place where the stack was originally positioned. From here the roofer can easily reach the next stack and continue shingling.

T-locks are a different situation. The best speed can be obtained with the minimum effort by taking four, five or maybe six courses across the roof at one time. In Figure 9-2 the roofer is taking six courses across. Most roofers can only handle four or five courses comfortably. Here the roofer has laid out four stacks of shingles with six shingles in each stack. Each stack is turned so that he can reach up and grab the shingle by the top, flip it around and slide it into place. If six shingles are too many for you, try five or four. Shingling across like this is faster because you don't have to move around as much. Also, in the summer you will have the advantage of being able to stay on cool shingles all the time.

### Shingle Selection

On reroofing consider carefully how the shingle you are installing matches the existing roof. The new shingles should fit nicely with the old shingles. Figure 9-3 shows a T-lock shingle being nailed on over a wood shingle. This particular T-lock gains seven inches with each course and the wood shingle gains only five with

T-lock shingles nailed over wood shingles
Figure 9-3

## Tips For Professional Roofers

3-tab shingles over a T-lock shingled roof
Figure 9-4

Uneven T-lock shingled roof
Figure 9-5

each course. Here a 3-tab shingle with a 5 inch exposure would have been the best choice. As you can see, the top of a T-lock falls in a different spot on the wood shingle on every course. There will be soft spots, dips and ridges all across the roof. A 3-tab shingle, installed using the butt-up method would have lasted much longer. Also, this roof will be very vulnerable to hail damage. The use of 15 pound felt underlay over the wood shingles would not improve the durability of the roof.

Figure 9-4 shows another example of a poor shingle selection. The roof below is a T-lock shingle roof which should never be covered up with a 3-tab shingle. If the T-locks can not be removed, then new T-locks should be applied because their basket weave appearance has the ability to cover up an unevenness in the roof. Here again, the use of a 15 pound felt underlay would not help the situation.

In Figure 9-5 the sun helps to demonstrate the unevenness of a T-lock shingled roof. Where the shingles lock together there is an area that has four layers directly above an area that has only two layers. This is the reason that a T-lock roof is impossible to cover up with 3-tab shingles and still expect a smooth roof. While another T-lock shingle would conceal most of this unevenness, a wood shingle or a heavy asphalt strip shingle with a shake appearance would make a much better roof. If a 3-tab shingle is wanted by the homeowner then you should convince him that the old shingles should be removed for a smooth finished roof. Prior warning should always be given because if you wait until the roof is finished he will think that you are only making excuses for poor workmanship.

Ribbon courses are sometimes called shadow courses as in Figure 9-6 and can be a nice feature in a wood roof, but if you reroof with asphalt shingles they must be removed. Some roofers cut off the exposed portion of the shadow course but this does not completely remove the problem. The only sure way, and usually the easiest, is to pull the top shingle of the shadow course out with your hands. This shingle is normally nailed on with the same length nails, (3d wood shingle nail) as the rest of the roof. Therefore the nails will not penetrate the roof deck as much, making the shingles easier to pull out. If you reroof with asphalt shingles, the

Roof with a serrated pattern
Figure 9-6

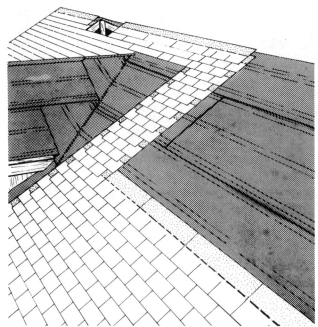

Using felt to keep shingles in line
Figure 9-7

shadow course will only appear to be an unsightly hump instead.

In the center of the roof in Figure 9-6 there is a section that has been roofed with a serrated (staggered) pattern. Here again the application of a 3-tab shingle would produce a very uneven surface. If low cost is essential and you must reroof with asphalt shingles, then T-locks would make a smoother roof. However, wood shingles or a heavy strip shingle would be much better. Again, the use of a felt underlay would only be a waste of time and money.

### Felt Underlay

Felt underlay must be put on as neat and straight as possible. This is sometimes called "capping" or "drying in". Do not wrinkle or buckle the felt as this will show up in the finished roof. If you keep it straight the felt can help you keep the shingles straight. Note Figure 9-7. Experienced roofers agree that one layer of 15 pound felt will make the best roof. Some architects and builders believe in the use of two layers of 15 pound felt or one layer of 30 pound felt. What they don't know is that usually the roofer has a difficult time trying to keep all of the buckles out of the felt. Producing one smooth layer is sometimes difficult so you know that two smooth layers must be twice as difficult. A 30 pound layer is also difficult to lay without any buckles. Some federal agencies require, in addition to the 15 pound felt underlay, a heavy layer of roofing paper applied to the overhang and extending eighteen inches past the inside wall. This is sometimes 30 pound or 45 pound felt and it usually does more harm than good, especially in cold weather. When the temperature gets warmer, this heavy felt will buckle, forcing the shingles up into a buckle also. The first hail storm that comes along will damage the buckled shingles because they have no direct support. Also, if any one should step on these buckled shingles, the buckle will move slightly and tear away from the nails. Any time that you are laying a double coverage asphalt shingle, there is no need for a heavier underlay than one layer of 15 pound felt, providing the shingles are applied correctly.

Correct way to waterproof a valley
Figure 9-8

Figure 9-8 shows the correct way to waterproof a valley. The felt underlay should always be watertight to prevent a leak in the event that you make a mistake somewhere on the roof. If a "freezeback" forces moisture under the shingles, the melting snow may never result in a leak if the felt is smooth and watertight. Be sure that you don't place any nails directly in the valley when nailing the felt.

### Flashing Around Vents

Vent flashings are often handled incorrectly, especially with T-lock shingles. Each vent presents a different problem because the shingle always seems to be in a different position in relation to the vent. Figures 9-9, 9-10 and 9-11 show almost the same situation. However, all three vents were handled differently. The main problem is that the T-lock shingle is long and interlocks.

*Tips For Professional Roofers*

Flashing installed over a whole shingle
Figure 9-9

Shingle installed on top of flashing
Figure 9-10

In Figure 9-9 the flashing was installed over a whole shingle and an extra piece of shingle was slipped in over the top of the flashing to make it watertight. Had the whole shingle been installed on top of the flashing, it would have been similar to the situation in Figure 9-10. This would leave too much shingle around the bottom of the flashing. Since T-locks are big shingles, the best way to handle this situation is to install the whole shingle on top of the flashing and then trim off the tab. Then from another shingle cut off a portion of shingle that includes the tab plus another two inches. Slip this shingle up under the bottom of the vent flashing. This will make the vent watertight and much neater. See Figure 9-11. The particular flashing is made of aluminum and must be nailed with aluminum nails rather than galvanized nails. Also note in Figure 9-11 that the shingle has been trimmed back from the flashing about one half inch so that dirt and granules will wash free of the flashing.

Shingling around a vent flashing with 3-tabs is quite simple compared to T-locks. The bottom of the flashing should always be left open and the rest of the flashing shingled over. Generally the bottom of the first shingle that goes over the flashing should be in line with the bottom side of the vent pipe. This will prevent water from going under the flashing at an angle and leaking through the roof around the pipe. Sometimes a cut out or bond line of this first shingle will be even with the edge of the flashing. This can result in a leak. To prevent this you can either slip a piece of shingle in under the bond line or apply plastic cement under the metal flashing. Always make clean, neat cuts around the vent flashing. Leave about a one half inch gap around the flashing so that dirt and granules will wash free. A tight cut will collect dirt and granules which will eventually cause the metal to rust. This can cause water to detour back under the shingles and leak into the sheathing. Don't use any more nails than necessary to hold the flashing in place.

Correctly installed aluminum flashing
Figure 9-11

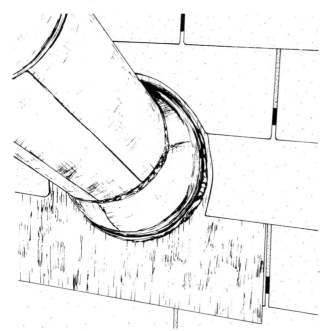

Flashing on a 3-tab roof
Figure 9-12

Incorrect installation of flashing on a shake roof
Figure 9-13

Flashing lying under the course it interrupts
Figure 9-14

Flashing on smooth wood shingles
Figure 9-15

Figure 9-13 is a good example of how a flashing on a shake roof should *not* be installed. The flashing should lie under the course that it interrupts, as in Figure 9-14. With these small vent flashings you should lay a double layer of 30 pound felt over the top half of the flashing. Be very careful with the joints in the shingles above the flashing as this is where most leaks will occur. On shake roofs it is very difficult to make repairs later. Do all that you can to prevent leaks while you are installing the roof.

On smooth wood shingles the vent flashings are easy to install. Note Figure 9-15. Be careful where you position the joints in the shingles above the flashing. Directly above the vent pipe in Figure 9-15 a narrow shingle was applied over the flashing to give support to the whole shingle that covers the top of the flashing.

On reroofs, whenever possible, raise the old flashing and shingle around it as though it were a new roof. However, sometimes this is not possible so you must cut your shingles around

*Tips For Professional Roofers*

Cutting close to the shingle
Figure 9-16

the flashing and seal it with plastic cement. Make the cuts as close to the vent as you can, as in Figure 9-16. Before you start sealing the vent, raise the shingles, as in Figure 9-17, and thoroughly seal under the shingle. Note the arrow in Figure 9-17 that is pointing to a small slot. Water will run in this slot and result in a leak. To prevent this the small portion of shingle must also be raised and sealed. While you are sealing under the shingles, seal the joint in the two shingles as shown in Figure 9-18. Now you can lay the shingle down and press it into the plastic cement. Finally, seal around this shingle. Be careful that you don't leave any small holes unsealed. After smoothing out the plastic cement, spray the vent with aluminum paint. This will improve the appearance of the vent, prolong the life of the plastic cement and serve as a good primer if the vent is painted later.

Extreme care should be used when sealing around vents as this is where many leaks occur on reroofs. On all types of asphalt shingles you must raise some part of a shingle and double seal. Water will run under these shingles at an angle and cause a leak. When you reroof with wood shingles or shakes you must always install new flashings. Plastic cement applied to any wood is only temporary because the wood will soak up water and release the plastic cement.

Usually you can raise the old flashing and shingle around it properly. Figure 9-19 shows a typical reroofing situation. The old vent is nailed down thoroughly. With a nail bar these nails could be removed easily. Then raise the flashing and let the old shingles lay back down on the roof. In some cases you will have to remove the old plastic cement from around the collar with a hatchet. Next shingle around the vent as though it was a new flashing. Make sure that there are no old nail holes in the flashing to cause leaks. Note in Figure 9-20 that the tab was cut from the shingle that was applied to the top of the flashing and a longer tab was inserted up under

Sealing the shingle
Figure 9-17

Sealing a joint between two shingles
Figure 9-18

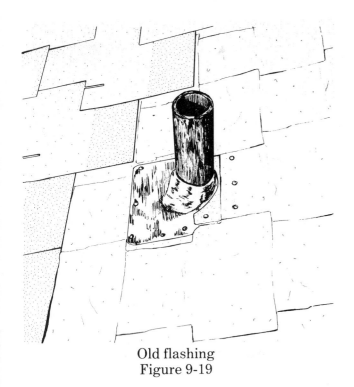

Old flashing
Figure 9-19

Tab applied to flashing
Figure 9-20

the flashing. This is sometimes necessary on a T-lock roof. This vent can now be sealed around the collar and spray painted.

There is no way to simply shingle around a flat attic vent when reroofing (Figure 9-21) and seal it with plastic cement so as to make it watertight. Most roofers will try anyway. It usually takes more time to try sealing a vent like this than it would to remove the nails, lift the vent and shingle in the vent properly. Figure 9-22 shows a typical situation that roofers encounter. Using a nail pry bar it only takes about thirty seconds to lift this vent. Then you can shingle around it much easier and will not have to come back to seal it later. Figure 9-23 shows that this produces a much more professional looking job than merely shingling a vent without lifting it.

Occasionally on a commercial roof you will encounter a large vent something like outlined in Figure 9-24. You will have to lay several shingles up along the side of this type of vent. Always leave a water trough about 1½ inches wide against the vent. Any whole shingles must have the top corner dubbed or cut just as in a valley. Otherwise water will run against the shingles and flow back along the top of the shingles until an opening is found. The extra precaution of applying plastic cement to the edge of the metal under the shingles is appropriate in many cases. When nailing the shingles, try to keep nails back a little from the metal.

Step Flashing

When you install metal step flashing on a shingle roof you should always use an individual piece for each course of shingles. See Figures 9-25 and 9-26. Water will get along the side of and then under the shingles unless individual step shingles are installed to turn this water back. On 3-tab shingles the metal should be at least seven inches long so that it will have a two

Flat attic vent
Figure 9-21

*Tips For Professional Roofers*

Attic vent before reroofing
Figure 9-22

Attic vent after reroofing
Figure 9-23

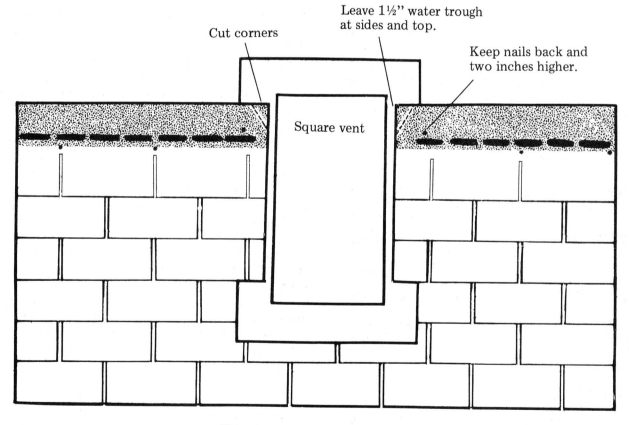

Shingling around a large vent
Figure 9-24

inch lap. It should extend up the wall at least two inches and out on the shingle about three inches. These are the minimum safe figures. On a T-lock shingle roof the flashing size and length will vary with the type of shingle and the individual situation. However, you should still have at least a two inch lap and the metal should extend up the wall at least two inches.

Flashing behind the fascia
Figure 9-25

Lifting the tabs to slip in flashing
Figure 9-26

Figure 9-25 shows a critical area. The roofer must loosen the fascia board and slip the step flashing up behind it. Sometimes this is difficult because the carpenters nailed the board up tight and didn't leave any space between it and the roof deck. To make the roof watertight, the flashing must go under the fascia.

Sometimes roofers will leave out the metal step flashing until the slope is shingled and then come back and install it. This is faster sometimes because the roofer, while he is shingling, can concentrate on shingling only. Then later he can devote all of his efforts to the flashing. To install the step flashing later he needs to only lift the tabs individually and slip in a piece of metal, as in Figure 9-26, and place one nail through the shingle at the top of the metal. The next piece of metal will cover this nail. You must start at the bottom and work your way to the top.

The metal step flashing in Figure 9-27 looks as though it is individual step flashing. It is not, however, because T-locks gain only about seven inches with each course. This metal is about sixteen inches long and is placed under the half shingle, on top of another shingle. To be watertight, a ten inch piece should be laid under the whole shingle, as in Figure 9-28. Then another ten inch piece of step flashing can be laid on top of the whole shingle, overlapping the other metal about two inches. Now the half shingle can be laid. The long piece of metal in Figure 9-27 could leak if enough water got under the half shingle at the top. The water would probably run on under the shingles before it got down to the end of that particular step flashing. While you are installing the shingles it will be very easy to install the proper flashing. If you have to come back after the siding has been applied to repair a leak it will be much more difficult.

Step flashing T-lock shingles
Figure 9-27

*Tips For Professional Roofers*

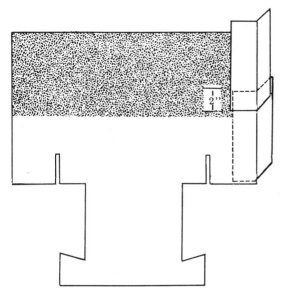

Watertight flashing for T-lock shingles
Figure 9-28

along the wall. With shakes this metal should be at least thirteen inches long and have four inches on the shingle and four inches up the wall. On smooth wood shingles you can use a 5 inch by 7 inch piece of metal, bending it near the center so that you have three inches on the shingle and two inches up the wall. These figures are minimums and may vary somewhat depending on the job situation and the builder's or architect's requirements. When you are installing the step flashing, make sure that each piece of metal is bent *less* than a full 90 degree angle. Then when you force each piece down to nail it in place it will hug firmly against the wall. This will make a much tighter flashing and be easier to seal later.

Flashing wood shingles
Figure 9-29

Finished flashing covered with aluminum paint
Figure 9-30

Roofs for house additions can be very difficult to make both watertight and neat looking. The hardest area to tie into is a stone or brick wall. If the builder or homeowner will not authorize the extra expense of sawing out the mortar joints and installing a metal counterflashing, the roofer will probably have to improvise the best that he can. If you are applying asphalt shingles, you may be able to seal the shingles with plastic cement. On a wood shingle roof or a shake roof you will have to flash the shingles as in Figure 9-29. To start out, install the step flashing as you lay the shingles

With the roof finished and all of the step flashing in place you are ready to seal the flashing. Begin by filling all the little holes where the metal flashing meets the mortar joints. This is easier if you use a caulking gun with plastic cement (tab cement). This eliminates voids and weak spots behind the flashing. Now apply a three inch wide strip of plastic cement to the wall, one inch down on the metal and about two inches up on the stone or brick wall. Be careful not to drop plastic cement on the shingles. Now take a three inch strip of membrane and apply it to the plastic cement, carefully rubbing it in so that it adheres completely. Over this apply a thin layer of plastic cement so that the membrane is thoroughly saturated and coated. To help

Roofer's Handbook

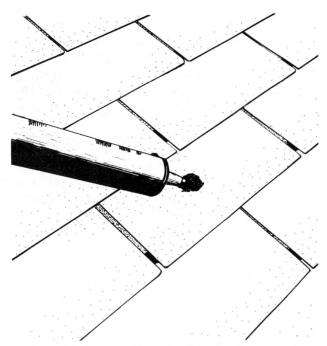

Repairing a small spot with plastic cement
Figure 9-31

Applying granules from chalk container
Figure 9-32

smooth out the flashing you should brush on a coat of good aluminum paint. This will prolong the life of the flashing. This type of job will not last nearly as long as a metal counterflashing but sometimes individual job requirements make it necessary. Figure 9-30 shows the finished flashing. When the aluminum paint is completely dry it can be painted over with a good latex paint the same color as the wall so that it will not be so obvious.

Spot Repairs

If hail damages a small area and you cannot match the shingle color, you can make spot repairs as in Figure 9-31. The best way to apply the plastic cement is with a caulking gun. The plastic cement (tab cement) can be purchased at most lumber yards. Apply the cement to the damaged spot and then sprinkle on some granules that match the roof. After the cement has been thoroughly covered, use your finger to press the granules into the cement. This will make a fairly good repair. Be careful when walking on the roof that you don't step in a freshly repaired spot and then on loose granules. This could be extremely hazardous. Granules of different colors can be purchased from many roofing suppliers or dealers. A handy container for storing and applying the granules is an old chalk container. Note Figure 9-32.

Replacing Individual Shingles
On a 3-tab shingle roof you need not replace the whole shingle if only one tab is damaged. First check to see if the shingles are the seal down type. If they are and the adhesive is factory applied, you must replace the shingles only when they are cool. Hot shingles can be damaged very easily. If the shingles are sealed down with plastic cement, the only time you can lift the tabs is when they are hot. Use a flat pry bar to lift the tabs around the damaged tab and remove the nails. You will need to lift the tab of the second shingle above and remove the nails that are holding the top of the damaged tab also.

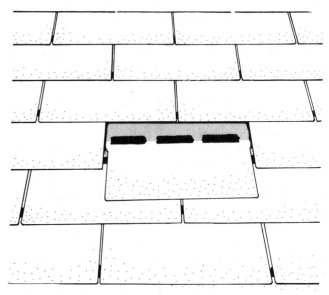

Replacing a tab
Figure 9-33

## Tips For Professional Roofers

Now hold these shingles up and cut the damaged tab free, using a utility knife with a hook blade. Replace the tab with a new tab and renail the shingles. To be on the safe side you should apply plastic cement to the old nail holes and the cut joints. Always apply a dab of plastic cement under the new tab as it will be limber and likely to be blown up by the wind.

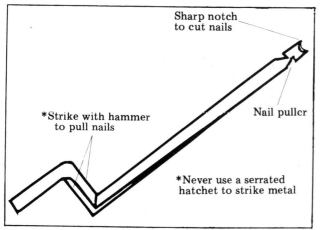

Nail cutter and puller
Figure 9-34

Replaced shingle held by nailed strip
Figure 9-35

Tile, slate and asbestos shingles are difficult to replace because you cannot raise the shingles to remove the nails. You can buy a tool, like that in Figure 9-34, and pull the nails by tapping up with a hammer on the base or tapping down on the handle. You can make your own nail cutter and puller from a piece of flat metal such as a car leaf spring. Sharpen one end into a slight V shape. By tapping on the unsharpened end you can loosen the nails and then pull the shingle free. To install a new shingle, cut a 1 inch to 2 inch wide strip of heavy galvanized metal or copper. Nail the strip into position so that when you slip the new shingle in and bend the bottom end of the strip it will hold the shingle in place. See Figure 9-35.

Wood and shake shingles can sometimes be removed by pulling them out by hand. Just lift the shingle up slightly, wiggle it around and pull it out. If wood shingles are old and brittle you may have to split them with the hatchet and then bend the nails over with a flat nail bar. To replace this shingle, just split a new shingle the right width and slide it up into place. You can nail wood shingles and shake shingles by placing the new shingle within a half inch of where it should be and then driving the two nails at an angle just below the butts of the course above, as in Figure 9-36. Now tap on the butt of the new shingle and drive it up in line with the other shingles. This will straighten the nails and

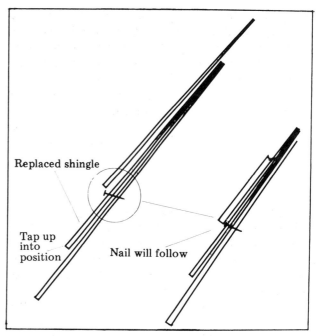

Hiding the nails
Figure 9-36

also move the heads under the course above.

To replace T-locks, lift up the tabs on each side of and directly above the damaged shingle. Remove the four nails as indicated in Figure 9-37. Then reach down and grab the damaged shingle at the tab, as in Figure 9-38, and slide it out. After slipping in the new shingle and renailing it, relock the shingles into place. Be very careful to bend the tab gently, as in Figure 9-39. If you bend the old shingles too much they

# Roofers Handbook

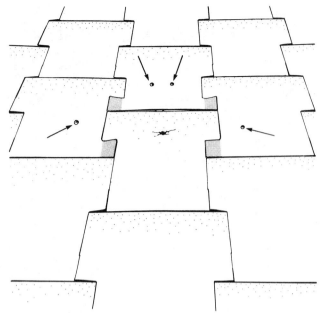

Removing nails to replace damaged T-lock
Figure 9-37

Sliding out the damaged shingle
Figure 9-38

Sliding in the new shingle
Figure 9-39

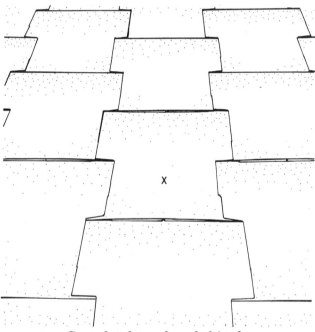

Completely replaced shingle
Figure 9-40

may break. With some types of T-locks you may have to tear off part of the ear portion of the shingle to get it back into position. T-locks should be replaced only when the shingles are warm. The completed replacement should not be noticeable. See Figure 9-40.

## Metal Edging

Proper installation of the metal edging is the first step in applying a good roof. Before you can apply edging you must get the edge ready. In Figure 9-41 the wood shingles have been trimmed back. This can be done with your hand or by splitting the shingles with a hatchet. At the bottom edge you will need to cut the shingles off with a sharp utility knife or an electric hand saw. Sometimes you may be instructed to trim the edges back far enough so that a 1x4 or a 1x6 can be installed on the roof edge. This really will not make the roof any better.

When you nail on the metal edging, make

*Tips For Professional Roofers*

Trimming back wood shingles
Figure 9-41

Nailing metal edging
Figure 9-42

Leaving the existing metal edging
Figure 9-43

shingles on a hip roof are very old, the metal ridge should be left on. Cut the metal edging as in Figure 9-44 and install it so that the old roof and ridge are concealed. In this case, when you trim back the old shingles you will need to trim back the metal ridge also.

On asphalt shingle roofs you should always trim the old shingles back and install metal edging over the edges. See Figure 9-45. Here

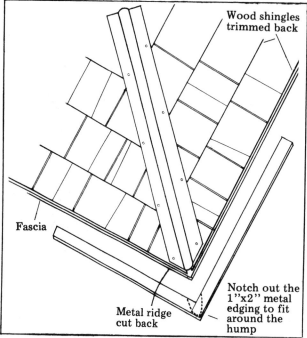

Cutting metal edging on hip wood roof
Figure 9-44

sure that you place the nails where there is some support under the metal, as in Figure 9-42. Nail through the thickest part of the old roof, usually the butt of the shingles. If you nail elsewhere, the metal will pull down and buckle the exposed part of the edging. On a wood roof with an existing metal ridge you should probably leave the old ridge on. The extra height can be used to keep the metal straight. This is because wood shingles are tapered and are not nearly as thick at the top as they are eight inches down, at the bottom of the last course. See Figure 9-43. If the also, nail through the thicker part of the shingles so that the metal does not buckle. After trimming the shingles you may need to slide the blade of your hatchet along the wood to knock off any heavy accumulation of paint that would prevent the metal from fitting snug against the wood.

Installing metal edges on an asphalt shingled roof
Figure 9-45

## Trimming Rakes

Precise trimming of the rakes after the shingles have been installed is the key to a nice job. This finish cut will be exposed for about fifteen years, so use extreme care. The new shingles at the ridge should not be bent over onto the other slope. This would make it difficult to determine exactly where the cut should be. Instead of measuring the cut with a tape measure, slip your index finger in under the shingles where the ¾ inch overhang will be and use the width of your finger as a gauge. Most index fingers are about ¾ inch wide but you can vary the estimate depending upon how big your finger is. At the top make a cut in the shingle about one inch long and drop the knotted end of a chalk line in the cut. Reel off the chalk line until you reach the bottom. Then, after checking with your finger and marking the shingle, snap the line. When cutting these shingles you must always start at the top and work your way to the bottom. To trim the right rake, as in Figure 9-46, start cutting at the top of the shingle and cut down. This is the natural side for a right handed roofer. To trim the shingles on the left rake, start at the top but you cut from the bottom of the shingle to the top, as in Figure 9-47. While cutting a shingle, use your left hand to hold the shingle and lean the knife over slightly away from the roof. This will make shingle cutting much easier. Pulling the knife too slow will make the cutting more difficult. If the temperature is high you may have to make short quick

Trimming the right rake
Figure 9-46

Trimming the left rake
Figure 9-47

*Tips For Professional Roofers*

Tearing off a roof
Figure 9-48

cuts to keep the hook blade from hanging up. If the shingles are hot, chalk a line on the rake as usual and then wait until later to do the trimming. Sometimes you can cover the shingles with a couple of boards or shingle wrappers until they cool. A sharp hook blade will make your job easier. Change blades frequently and store the dull blades inside of your utility knife until you have time to sharpen them. They sharpen very easily on an electric grinding wheel.

## Tear Off Tips

No one enjoys tearing off a roof. However, using the right tools and the right methods will make it much easier. On wood roof tear offs use a three foot crow bar. Place the hook between the sheathing boards and the shingles. With an upward sweep you can rip the shingles off easily and quickly. After a little practice with this method you can remove several squares in one hour. Always start at the top and work your way down the slope, taking off two or three courses at a time. See Figure 9-48. Be aware of the weather as you tear off a wood roof. If the weather suddenly changes, it will take more time to waterproof a wood shingle roof than an asphalt shingle roof.

The best method for removing asphalt shingles depends on how old the shingles are. In Figure 9-49 the shingles are old and brittle. Here the best method is to simply start at the top and use a flat spade to pry the shingles and nails up. If the shingles are new enough so that they don't fall apart, use the flat spade to pop up the nails, working your way across or stair stepping down the same way the shingles were installed. Keep the shingles intact if possible and stack them up in bundle-like stacks. Remove the stacks from the roof as necessary. Of course, this isn't possible on a steep roof. But on a low pitch roof use of this orderly method will save time and effort. Have the man with the shovel work across or at an angle. Another man should follow, removing and stacking the shingles and throwing the stacks into a truck or trailer. Shingles that fall apart should be removed with a scoop shovel.

If you are removing more than one layer of

99

Removing old asphalt shingles
Figure 9-49

roof, use the same method and remove only one roof at a time. If you try to tear off two roofs at the same time you will encounter too many nails, making the tear off very difficult. When you remove the first layer of roof you should try to remove all the nails in that layer, making the next layer much easier to remove.

## Roll Roofing Tips

Ninety pound roll roofing makes an easy and economical roof on slopes with as little as 1 in 12 pitch. Since it is a single coverage roof, you must make sure that the old roof is smooth and that the roll roofing goes down free of buckles. A roll is 36 inches wide, 36 feet long, contains 108 square feet, and weighs 90 pounds. The laps should be sealed together with plastic cement and nailed about every two inches. When you reroof with 90 pound roof over a 90 pound roof, start out at the bottom with a different width strip than the old roof used. Usually the old roof was started with a full 36 inch strip. You should then start with a half strip so that your overlap will fall in a different area, as in Figure 9-50. If you allow the overlap to occur in the same place, a hump will develop.

Temperature is very important when applying roll roofing. The roof in Figure 9-51 was applied when the temperature was warm so the material was simply rolled out and nailed in place. Plastic cement was liberally applied to the two inch lap each time before a new roll was started. Be careful that you don't apply too much plastic cement as it will squeeze out onto the roof. But apply enough to seal the two pieces together. If the weather is cold or even cool you should cut the material up into 12 foot lengths and allow these to warm in the sun before application. This will eliminate buckling.

If the flat area ties into a shingle roof that is not to be reroofed, remove the nails in the first two courses and slip the roll roofing up under the shingles at least twelve inches. If the shingle roof is going to be reroofed, you can just turn the

Overlapping roll roofing
Figure 9-50

Applying roll roofing
Figure 9-51

*Tips For Professional Roofers*

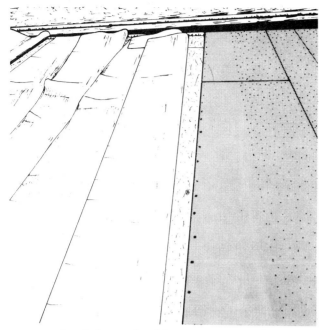

Applying a double coverage roof
Figure 9-52

Applying adhesive to double coverage roof
Figure 9-53

material up on the shingles. But be sure that the shingles are applied before it rains. Any soft spots or voids where you turn up the roll roofing on the shingles should have a double layer of roll roofing. Keep in mind the heavy rain and snow that this roof will have to endure, especially if a shingled area drains onto the 90 pound roll roof. Like all other roofs, the edge of the old roof should be trimmed back and new metal edging installed. The only extra precaution needed is plastic cement applied to the metal all the way around the slope. Make sure you have plastic cement under the material anywhere you place nails.

On low pitch roofs (1 in 12 minimum), a double coverage roof can be applied. This double coverage roll uses a 19 inch selvage edge sheet that comes in 36 foot rolls, 36 inches wide, with 108 square feet. One roll will only cover one half of a square because double coverage is required. In Figure 9-52 the entire roof slope has been covered, starting at the bottom and lapping each layer 19 inches. The material is nailed at the top edge every twelve inches. Then, starting at the top, the entire roof was folded back, one sheet at a time. The roofer then applied a cold application adhesive designed for this type of roof to the 19 inch lap portion of the strip. See Figure 9-53. The adhesive is being brushed on with a throw-away roofing brush at the rate of 1½ gallon per square. After each strip is thoroughly coated, getting as close to the lap line as possible, it can then be flopped over and rubbed down completely. This material, like all other roll roofing, should not be applied when cold. If in a sunny area, it can be nailed on and permitted to warm in the sun for several hours before sealing.

Don't be reluctant to recommend roll roofing. It can be purchased in most of the same colors as asphalt shingles and will make an excellent roof if it is applied correctly.

# Chapter 10
# Handling Leaks

Repairing roof leaks can be very difficult. It is usually futile to crawl up in the attic and search for where the water is coming through the roof. In most cases the water will run under the shingles for some distance, to one side or the other, and then down until it finds a way through the roof deck. If there are two roofs on the house, the new roof may leak for some time before it appears in the house. There are many cases where an addition was built on a house and the first rain disclosed a leak in the new addition. Of course, this will seem as though the addition is leaking. However, the wrong roofer may get the call-back. Such was the case in Figure 10-1. The slope on the right is the new addition. The roofer apparently did a good job of tying the two roofs together at the valley. However, a leak developed in the addition. The leak was later discovered to be coming from the flat attic vent on the old roof, up about ten feet from the valley. Unfortunately, someone tried to repair the leak by applying plastic cement to the edge of the shingles in the valley. This never stops a leak and may even cause one. The bottom edge of a shingle should never be sealed off because the shingle then will trap the water that runs under the edges and this back-up may cause a leak. See Figure 10-2.

The best way to determine if a vent is leaking is to remove a couple of shingles directly below the vent after a rain. If a vent is leaking there will probably be some moisture between the shingles and the sheathing. This may make it appear that the valley is leaking. This was the case in Figure 10-1. The vent here was sealed properly and the leak was stopped. Had the vent been properly shingled around, as in Figure

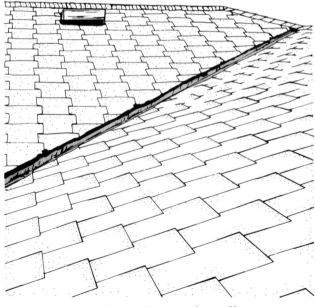

Addition tied on at the valley
Figure 10-1

*Handling Leaks*

Trying to repair a leak by applying plastic cement to the valley
Figure 10-2

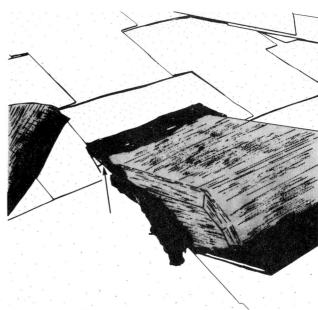

Improper sealing of vent
Figure 10-3

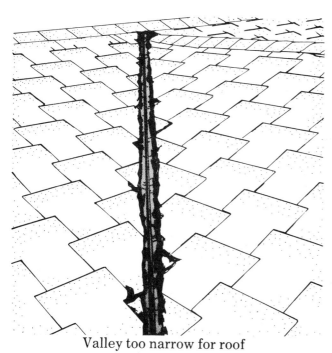

Valley too narrow for roof
Figure 10-4

Recut valley trough
Figure 10-5

9-23 or had it been sealed properly, the leak would never have developed. This vent probably leaked ever since the main part of the house was reroofed but never appeared until the addition was built, giving the water an opportunity to run into the house. Any time a roofer installs shingles on an addition he should check the entire area above his work. This type of leak occurs quite often. If he can discover the leak before he leaves he may not be called back unnecessarily. The leak problem in Figure 10-3 is similar to the vent leak in Figure 10-1. When the roofer sealed this flat vent he simply trowled the plastic cement around the vent without raising the shingles and sealing under them. The arrow is pointing where the leak occurred.

Valley Leaks
Leaks often occur in valleys and can develop from many different causes. The valley in

103

Dubbing the corners
Figure 10-6

Completely restored valley
Figure 10-7

Figure 10-4 was cut too narrow and the corners were not dubbed. Since the roof was new it was easy to repair. The first step was to recut the valley trough, as in Figure 10-5. This allowed heavy rains to run off the roof faster. Then the corners were dubbed. This was probably the main reason that the valley leaked. See Figure 10-6. After the plastic cement was scraped up, aluminum paint was applied and the valley was back in nearly new condition. See Figure 10-7. When the corners were dubbed, there were a few nails that were too close to the valley. These were removed and the holes sealed with plastic cement. The entire repair took only a few minutes but it would have taken less time for the original roofer to do it right.

Dubbing the corners in the valley cannot be over emphasized as this is the reason for most roof leaks. Figure 10-8 illustrates a wood roof where the corners were not dubbed. Wood shingles act the same as asphalt shingles when water runs down the valley. Also, the valley material here should have been W shaped formed valley because one slope was steeper than the other. If you suspect a leak in a wood shingle valley, look at the edges carefully to determine if the corners are dubbed. To trim these shingles, use a screwdriver to chip off the corners.

The leak problem illustrated in Figure 10-9 was made worse because the corners were not dubbed. The most serious problem was that the valley was not cut wide enough. Also, the valley trough was off center, causing water to run strongly against the edge of the shingles. Even if the corners were dubbed, there was enough water running under the edges to cause a leak. This valley could have been repaired had it not been for the large amount of plastic cement that was on the valley metal. All that was needed was to recut the valley on the problem side, dub the corners, raise the shingles and apply a ½ inch strip of plastic cement about six inches back from the edge. Finally, to be sure of a positive

Wood roof corners that need dubbing
Figure 10-8

*Handling Leaks*

Leaking valley
Figure 10-9

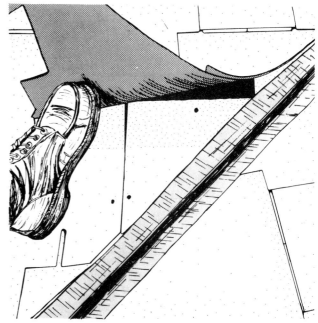

Improper installation of shingles
Figure 10-10

seal, the shingles should have been pressed gently into the cement. Instead, the valley had to be removed and a W shaped formed metal valley installed.

The leak problem in Figure 10-10 was not just a shingle corner that wasn't dubbed; it was a shingle that should never have been installed.

Sharp corners are not the only causes of leaks in a valley. Figure 10-11 illustrates a leak - prone situation. Water that comes rushing off the valley will hit a bond line of the shingles below. Then the water can run across the shingle under the tab, staying above the factory adhesive. If there is a joint in the shingles under this tab, a leak is almost certain. Sometimes the factory adhesive is continuous or the area between adhesive strips dams up from dirt and granules, permitting the water to continue flowing horizontally instead of running out from under the tab. This is not a common leak but it has happened.

A full lace valley can leak and will be harder to repair. The valley in Figure 10-12 was leaking but the water was showing up under the flat deck below the valley. Of course, this is where all of the repairs were made. Apparently no one suspected that the leak was originating in the valley. In Figure 10-13 you can see the reason that the full lace valley failed. The shingles were not slid up under every other course and some were slid up under a shingle two or three courses up. This let the water run in under a bond line and on down under the shingles in the valley. The water traveled under the shingles to the flat roof and then leaked through the roof at the flashing where the two roofs joined. This valley was not repaired because of the age of the shingles. Instead the house was reroofed and a new valley was installed. If a valley like this has to be repaired, remove the shingles on both sides of the valley and install a new valley.

A leak prone situation
Figure 10-11

105

Leaking full lace valley
Figure 10-12

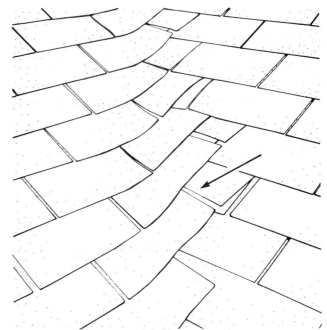

Improperly constructed full lace valley
Figure 10-13

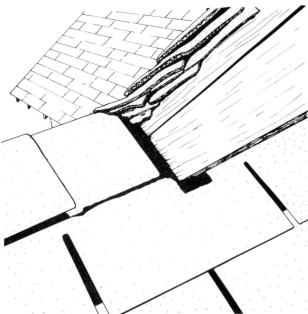

Asphalt shingles around a barge rafter
Figure 10-14

Wood shingles around a barge rafter
Figure 10-15

### Vent And Flashing Leaks

Most roof leaks occur where the shingles have been cut around something such as a vent, valley, fireplace or chimney or where a barge rafter comes down to the roof, as in Figure 10-14. These areas should be sealed with plastic cement on a reroof job and flashed with metal on a new roof. Figure 10-15 shows the same situation on a wood shingle roof. This joint can not be sealed with plastic cement as the cement would not last very long. Instead, it must be flashed with galvanized metal. The most common leak on a wood shingle roof occurs where joints or splits line up in three shingles in three adjacent courses as in Figure 10-16. This roof can be repaired easily by slipping a piece of flat galvanized metal up under the cracks. The metal is referred to as a tin shingle and can either be cut from flashing metal or purchased from a lumber dealer. These tin shingles will slide out in a short while and the leaking will start again if you don't bend the bottom corners

*Handling Leaks*

Applying a tin shingle to a wood roof
Figure 10-16

down as was done here. The sharp corners will dig into the wood shingles, keeping the tin shingle in place. Do not nail the tin shingles in place. The nails will gradually work loose.

Vents are often a good place to look for a leak. They are frequently sealed improperly or shingled incorrectly, as in Figure 10-17. The two tabs with the X mark should have been under the vent flashing instead of on top. Water must be able to wash free of the flashing. Here the tabs collected dirt and granules and this eventually led to rusting of the metal flashing. The dirt and granules also diverted the rain water under the shingles and probably made the leak worse. At the top of these tabs (see the arrow) was a sharp corner that also diverted water under the shingles. Had it been necessary to place the tab in this position, the top corners should have been trimmed off and the bottom part of the shingle should have been straight down so that water could wash away from the flashing. This method could be used to repair

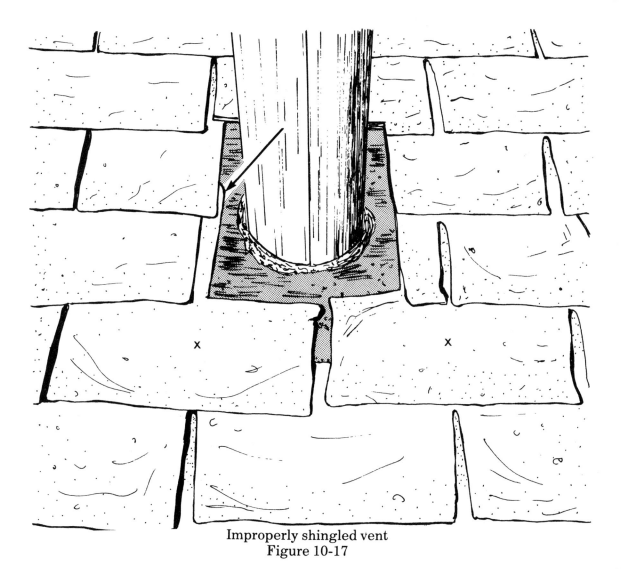

Improperly shingled vent
Figure 10-17

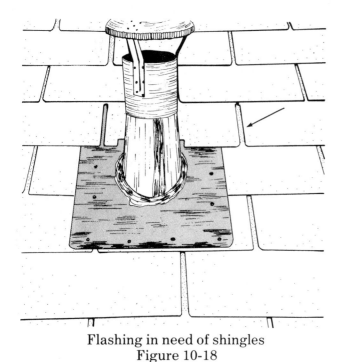

Flashing in need of shingles
Figure 10-18

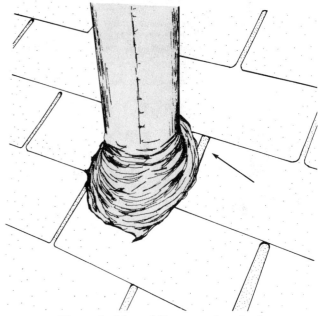

Unsealed bond line causing leak
Figure 10-19

this vent or the two tabs could be removed and placed under the flashing.

Sometimes a leak will occur because there are not enough shingles around a flashing. See Figure 10-18. The top edge was under the shingles but the water was coming in from the sides. Note the arrow. This flashing should have had one more row of shingles over it to be watertight. The vent in Figure 10-19 was leaking also and was in the same area of the roof as the vent in Figure 10-18. The arrow is pointing to a

short bond line next to the vent. This bond line was not sealed when the vent was sealed with plastic cement. To repair the vent in Figure 10-19, raise the tab and seal around the top of the vent. Also seal the short bond line. These two vents are the middle and lower vents in Figure 10-20. The leak showed up in the flat roof at the bottom of the shingle roof. It appeared to be a single leak. But only when both vents were repaired did the leak stop. Remember that water will sometimes run under the shingles for

Location of the two leaking vents
Figure 10-20

Typical Leak situation
Figure 10-21

## Handling Leaks

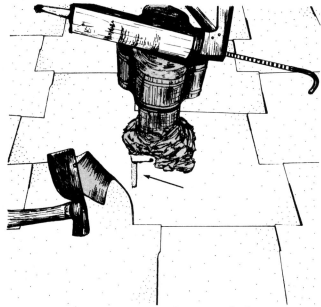

Raising tab to find exposed slot
Figure 10-22

Water leak under tab
Figure 10-23

twenty feet. So when you look for the cause of a leak, do not confine your search to a small area around where the water is showing up in the house.

Repairing leaks around a vent on a T-lock roof requires a little more attention because of the way the shingles lock together. The vent in Figure 10-21 is a typical leak situation where the water ran in the lock portion of the shingle and leaked under the plastic cement. When searching for the leak, just raise the tab and you can see the slot clearly. Note Figure 10-22. In Figure 10-23 the lock was never hooked in position but the water still ran under the tab and leaked. To repair a vent that has been improperly sealed on a T-lock roof, raise the tab and thoroughly seal off the slot. Then lay the tab back down but *do not relock* the shingle as this will disturb the plastic cement. Then the vent will probably continue leaking. After you lay the tab back

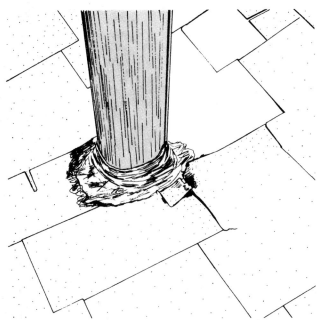

Plastic cement and aluminum paint
applied to base
Figure 10-24

Too much plastic cement
Figure 10-25

A faulty joint
Figure 10-26

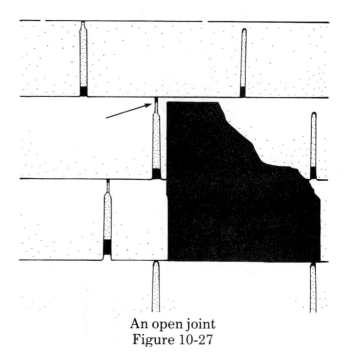

An open joint
Figure 10-27

down into the plastic cement, just press into place. Then apply more plastic cement to the shingle around the vent and spray with aluminum paint. See Figure 10-24. Sometimes too much plastic cement, if improperly used, will cause a leak. Note Figure 10-25. The vent was sealed like this by the heating contractor after the roofer applied the flashing. Of course, if it leaks, it will be the roofers fault. The heating man won't get the call-back. Plastic cement must not be applied to the bottom of the flashing and the bottom of the shingles if water is permitted to run under the shingles above. Water must have a way to run out or it will back up and result in a leak.

### Other Leaks

Occasionally a leak will occur out in the middle of a slope. In this case, look for a nail that is leaking, a joint in the shingles at a low spot, or a joint in the shingles where the exposure is over five inches. Note the circle in Figure 10-26. Figure 10-27 is a close-up of the problem. The dark shingle demonstrates how far the shingle underneath extends. The arrow is pointing to the leak where the shingle ends. You can actually see the old roof through the joint. Repair this by raising the tab and sealing the joint completely with plastic cement. As you can see, this roof must have been installed by an amateur. Look for other possible leaks in this same area.

Old fireplaces often leak because they never get repaired. This type of leak can show up about anywhere because water enters the top of the chimney (see Figure 10-28), runs down at random and comes out almost anywhere. This can be repaired by placing plastic cement, membrane and aluminum paint over the entire top of the chimney. For other leaks around the chimney, check the top of the counterflashing for loose joints. These can be sealed with a caulking gun using a butyl sealant.

Other items to look for around the chimney are debris trapping water on the back side or loose and improperly applied step flashing. Occasionally the counterflashing may leak in a

Old leaky chimney
Figure 10-28

*Handling Leaks*

Decoration nailed to roof
Figure 10-29

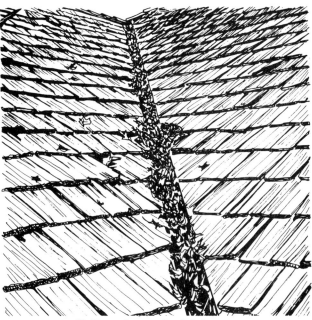

Valley clogged with leaves
Figure 10-30

blowing rain because it was not lapped enough.

Some roof leaks can be attributed directly to the homeowner as a result of decorations placed on the roof. These will probably be hard to find because the leaks are the result of small nail holes. Sometimes the decoration will cover a large area in the middle of the roof. See Figure 10-29. This is one leak that can be seen from the attic because the water falls straight through the shingles and roof deck. To repair these holes, raise the tabs and seal with plastic cement.

A common leak occurs in the valley when the flow of water is obstructed by a build-up or accumulation of leaves. Note Figure 10-30. The valleys should be kept clean by sweeping with a broom. Sometimes a leak will occur at the bottom of a valley when the house has no overhang and the gutter is stopped up, backing water up onto the roof. A roofer should check and clean the gutters when he is applying a new roof so that he will not be called back on a leak such as this.

# Chapter 11
# Wood Shingles and Shakes

Wood shingles and shakes can be applied by anyone reasonably skilled with simple tools, but highly professional, rapid application is an art that is mastered by relatively few. After you learn the fundamentals of wood roofing you can develop speed only through perseverence, practice and self discipline. An accomplished wood shingler moves like a robot in high gear yet doesn't sacrifice quality for quantity. This chapter has the fundamentals you need. The rest is up to you.

### Tools And Equipment

Proper tools and equipment are essential to professional wood roofing practice. The hatchet used is normally a lightweight lathers hatchet with a non-skid head. This head prevents the hatchet from slipping off of rough galvanized nails. The blade is used for cutting and splitting the shingles. Keep it sharp. Many roofers cut off part of the handle so that the hatchet has a better balance. The gauge is important for wood shingles. If the hatchet has no predrilled holes, drill your own holes or use the sliding gauge. A sliding gauge that slips slightly will result in misaligned shingles.

A nail stripper will help anyone obtain more speed with less effort. If you have not used a nail stripper, give yourself time to get accustomed to it. The stripper may seem to be a bother at first. Best results can be obtained from a stripper by not overloading it. Usually a small handful of nails at a time is enough because too many will overload the troughs, making it difficult for the nails to fall into place. Wear a nail bag so that you will have a supply of nails nearby at all times. As required drop a small handful into the stripper. Occasionally, as you are shingling, you may need to jiggle the nails so that they drop into the troughs. Keep the stripper angled down slightly towards the doors so that the nails fall into place naturally. Many roofers spray the nails with mineral oil to make them move more smoothly in the stripper. Keep the discharge doors lubricated and wear the stripper tight to your chest so that you can jerk the nails out easily. A hip pad will help keep you from slipping, protect your clothing and provide some cushion from hot, rough shingles. You can make a hip pad from a truck inner tube as described in Chapter 1 or purchase a thicker manufactured hip pad. If the stripper and hip pad can not be purchased locally, order either from the following companies: South Coast Shingle Company, 2220 East South Street, Long Beach, California 90805; Roofmaster Products Company, P. O. Box 63167, Los Angeles, California 90063 or McGuire-Nicholas Mfg. Company, 6223 Santa Monica Blvd., Los Angeles, California 90038. Be sure to specify whether you are left or right handed. The stripper comes in a nail size for wood shingles and a nail size for composition shingles.

The other tools that you need, such as a tape measure, chalk line, utility knife and tin snips, were described in Chapter 1. For cutting the valley and hip shingles you need a circular power saw, especially on shakes. You can usually cut wood shingles with a hatchet or a straight blade utility knife. Mechanical hoists are not as widely used with wood shakes and shingles because of their lighter weight. However, on some jobs they can be a tremendous advantage.

### Material Selection

Good application starts with selecting the right shingles for the job. A good wood shingle roof is never less than three layers thick. Consequently the exposure of any given shingle must be slightly less than ⅓ the total length. If the roof pitch is 4 inches per foot or more, allow a 5 inch exposure for 16 inch shingles, a 5½ inch exposure for 18 inch shingles and a 7½ inch exposure for 24 inch shingles. If the roof pitch is less than 4 inches per foot but not below 3 inches per foot, allow a 3¾ inch exposure for 16 inch shingles, a 4¼ inch exposure for 18 inch shingles and a 5¾ inch exposure for 24 inch shingles. If the roof pitch is less than 3 inches per foot, cedar shingles are not recommended. The figures above are the maximum recommended exposures for Number 1 Blue Label shingles. When applying Number 2 Red Label and Number 3 Black Label shingles, the exposures must be somewhat less, as recommended by the Red Cedar Shingle and Handsplit Shake Bureau. See Table 11-1.

| Pitch | No. 2 Red Label | | | No. 3 Black Label | | |
|---|---|---|---|---|---|---|
| | 16" | 18" | 24" | 16" | 18" | 24" |
| 3 in 12 to 4 in 12 | 3½" | 4" | 5½" | 3" | 3½" | 5" |
| 4 in 12 and steeper | 4" | 4½" | 6½" | 3½" | 4" | 5½" |

Table 11-1

The minimum recommended pitch for handsplit shakes is 4 inches per foot. The maximum exposures for 24 inch shakes is 10 inches and 7½ inches for 18 inch shakes. A superior 3-ply roof can be obtained at slight additional cost if the exposures are reduced to 7½ inches for 24 inch shakes and 5½ inches for 18 inch shakes. Note that the 24" by 3/8" shake should be applied at exposures not greater than 7½ inches on roof pitches of less than 8 inches per foot.

The nails are the next most important part of a wood shingle or shake roof. Use only rust resistant nails, either zinc coated or aluminum. Figure a little over 2 pounds per square for both shingles and shakes. Use 3 penny for 16" and 18" wood shingles and 4 penny nails for 24" wood shingles. For reroofs, use 5 penny nails for 16" and 18" shingles and 6 penny nails for 24" shingles. Never use bright, blue or steel wire nails as the bare iron is not compatible with the natural preservative in the cedar shingles. In less than ten years the shingles will begin to dissolve the nails.

The 6 penny nail is usually adequate for handsplit shakes but be sure that the nail penetrates at least ½ inch into the sheathing. Sometimes a 7 penny nail may be necessary. For shake reroofs, a 7 penny or 8 penny nail will probably be needed. Always use 8 penny nails for shake hip and ridge units.

Proper positioning of the nails is very important. Two nails are required per shingle regardless of the width. Always nail within ¾ inch (1 inch for shakes) of the side edge of the shingle. Nail high enough so that the nails are covered by the next course. On wood shingles with a 5 inch exposure, nail about 7 inches from the bottom edge of the shingle. On shakes with a 10 inch exposure, nail about 12 inches from the bottom edge. If you nail too high or too far in from the edge the shingles will tend to curl up. Many times the sheathing is spaced too far apart and the roofer must raise his nails to hit the sheathing boards. This will let the shingles turn up somewhat. The builder will probably blame the roofer or consider the shingles faulty when the fault actually rests with the tradesman who applied the sheathing. When nailing the shingles at the rake edge, try to place the nails as close to the rake as possible. You may have to nail through the fascia or trim to keep the shingles flat. Any type of wood that is left sticking out unnailed will eventually curl up. This is why proper nailing is so important. If the shingle splits while you are nailing it and the crack offsets the joint in the shingle below at least 1½ inch, place a nail on each side of the split. You can treat the split shingle as two shingles. If the crack does not offset the joint in the shingle below, remove the split shingle and apply another.

Reroofing with wood shingles and shakes over wood or asphalt shingles will make an excellent roof provided that new valley flashings and vent flashings are used. If you are reroofing over spaced sheathing, don't worry about hitting the decking with your nails. Nail the shingles in the normal position as though you

Roofers Handbook

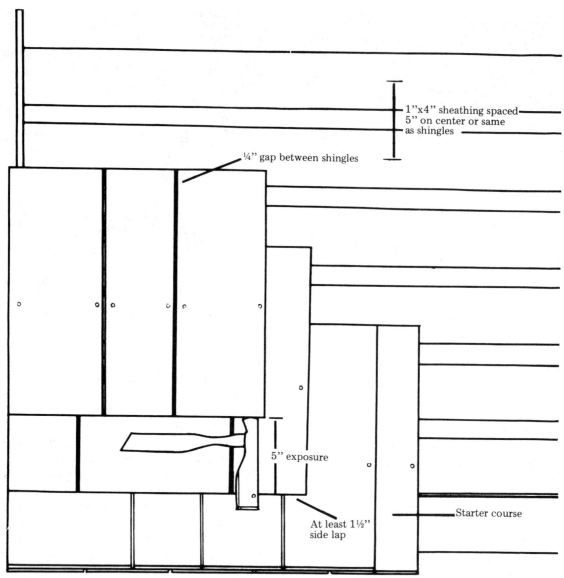

Wood shingle starting procedure & spaced sheathing
layout (16" shingles with 5" exposure)
Fig 11-1

were roofing over solid sheathing. Only if the old wood shingles are too thin to hold nails should you be careful to place nails in the decking. Reroofing with wood over wood in wet southern climates, especially in shaded areas, is not recommended as the new shingles may deteriorate prematurely. In this case, tear off the old roof before roofing.

### Ventilation

All roof attics must have proper ventilation. Lack of proper vents can be the cause of early roof failures. Check the house and advise the builder or homeowner what is needed. A gable roof should have a vent in each gable end and the vent should be as close to the ridge as possible. The net area for the openings should be 1/300th of the ceiling area of the rooms below the roof. For example, where the ceiling area is 1,200 square feet, the net ventilator area should be 4 square feet. Hip roofs should have air-inlet openings in the soffit area of the eaves and outlet openings at or near the peak. Inlet vent area and outlet vent area should each be 1/900th of the ceiling area.

### Estimating

Measure the roof length and width. Then multiply the length by the width and divide the total by 100. The result is the number of squares of shingles or shakes you will need to cover the slope using standard exposures. Don't forget to allow for double coursing (starter course) at the eaves. With wood shingles, one square will provide about 240 lineal feet of double course.

Starting in left hand corner
Figure 11-2

Gauging the overhang on a starter course
Figure 11-3

With shakes, one square will cover about 120 lineal feet. Also, for every 100 feet of valley allow one extra square of wood shingles and two squares of shakes. These figures assume that the ends cut off for valleys can not be used on hips and are wasted. If the roof is a hip roof with no valleys, figure the waste the same as for the valleys. If any cut off ends can be reused, figure the waste accordingly.

One bundle of factory produced ridge units will cover 16⅔ lineal feet for both wood shingles and shakes. On shake roofs you must figure about 1½ rolls of 30 pound felt 18 inches wide for each square of shakes at 10 inch weather exposure. Thirty pound felt interlay is not needed in snow-free areas when straight-split or taper-split shakes are applied at exposures less than one-third the total shake length (3-ply roof).

## Application Of Wood Shingles

Proper application starts with properly spaced sheathing. The sheathing should be spaced the same as the shingle exposure so that the roofer does not have to raise his nails too far. Properly spaced sheathing ensures a tighter and better roof. The measurement for spaced sheathing depends on the exposure of the shingles and is easy to figure out. Start with the second sheathing board as this is where the first row of shingles must be nailed. If the roof exposure is to be 5 inches, measure from the center of the second sheathing board and set the third board on a 5 inch center from the second board. Now measure the space between the two boards and make a wooden gauge to space the rest of the sheathing. Use 1x4 lumber for sheathing. This will take only three or four more 1x4 boards to a slope but will improve the durability of the roof significantly.

Figure 11-1 shows the procedure that is recommended for wood shingles. The shingle edges should not be crammed together. Leave a ¼ inch space to prevent buckling. Always leave a side-lap of at least 1½ inches between side laps in successive courses. Try to prevent laying two joints in line if they are separated by only one course of shingles. This way, if the one shingle between the two joints ever splits there will be no leak. At the eave or drip edge always allow a 1¼" to 1½" overhang. On gable roofs leave about a ¾ to 1" overhang at the rake edge. Many roofers use the first joint in their index finger to gauge this overhang. You can tack a temporary 1x2 board into place to guarantee a straight edge.

Right handers normally start in the left corner, as in Figure 11-2, and apply enough shingles so that they can sit down to shingle. Take a little extra time getting started because the rest of the roof will be gauged from your beginning course. If you start out crooked you will have to stop and straighten the courses out later. To gauge the starter course, use the method illustrated in Figure 11-3. Of course, hatchets vary somewhat, but generally this will give you about a 1¼ inch overhang. It's a good method to keep the shingles straight. Regardless of the pattern, the next row of shingles will lay straight and directly on top of the starter

Laying a serrated pattern
Figure 11-4

course. From there continue gauging the shingles as in Figure 11-1.

In Figure 11-4 the roofer is laying a staggered or serrated pattern. There are two different exposures in each row. Normally the offset should be only one inch. Starting at the bottom, the first row on top of the starter course is straight. The second row is where you start the staggered pattern. The "up" shingle is gauged 5 inches and the "down" shingle is gauged the same as in Figure 11-3, using the hatchet handle as a gauge. If the hatchet has a curved handle, leaving you nothing to gauge from, file a notch to gauge by. As you lay the stagger pattern you gauge each shingle up or down from the previous shingle in the course. You will need to get accustomed to this stagger pattern before it looks attractive to you.

Occasionally you may need a straight line at the drip edge to keep the starter course even. It is good practice to use a straight line if the edge is too crooked to gauge from or if the amount of overhang is too much to gauge easily. Many roofers try to avoid this method as it is somewhat slower than gauging. In Figure 11-5 you can see that the roofer is carefully aligning the starter course along the straight line. Here he is roofing above a mansard roof and needs more overhang than normal. To fasten the line, simply nail a shingle at each end in the desired position and tack a nail in the butt end of each shingle.

Aligning the starter course
Figure 11-5

Applying five rows at a time
Figure 11-6

Holding the shingle in place while
reaching for nails
Figure 11-7

In Figure 11-6 the roofer is taking 5 rows of shingles across at one time. Five rows is about as much as he can reach. Never apply more courses at once than you can reach comfortably. In Figure 11-7, after gauging the shingle, the roofer is holding it in place while reaching to the stripper for more nails. Then, as he places the nail on the shingle, his left hand holds the shingle in place while he drives the first nail. Here is where speed can be obtained. Both hands must work independently, sometimes at different tasks. In Figure 11-8 the roofer is driving down a nail with his right hand while his left hand is jerking more nails free of the stripper. Keep your eyes on the nail you are driving, not the nails you are pulling. Save time by glancing briefly at the loose shingles and then picking up and using a shingle of the right width. Don't waste time trying out different shingles. Many roofers, when taking 5 rows across at a time, always have two possible spots available to lay a shingle. This way, if they make a mistake and pick up a shingle that won't fit in one location, they may be able to fit it in the other spot. The roofer in the illustrations is making every move count, making automatic moves with deep concentration. In your effort to obtain more speed, don't forget to allow the 1½ inch sidelap on each shingle.

Figure 11-9 shows a roofer driving rust resistant staples with a staple gun. Always place the staples square with the shingle and in the same areas where you would use nails. While he is stapling the shingle with one hand, his other hand is reaching for another shingle. When using a staple gun, always wear a nail bag to carry extra staples. You can eliminate much of the weight and pull of the air hose by fastening it to your nail belt at a point about 5 or 6 feet from the gun. This will make the staple gun easier to handle by putting the weight on the belt instead of on the gun.

Driving the nail while reaching for more nails
Figure 11-8

## CERTIGRADE RED CEDAR SHINGLES

| Grade | Length | Thickness (At Butts) | No. Of Courses Per Bundle | Bdls./Cartons Per Square | Description |
|---|---|---|---|---|---|
| No. 1 Blue Label | 16" (fivex) 18" (perfections) 24" (royals) | .40" .45" .50" | 20/20 18/18 13/14 | 4 bdls. 4 bdls. 4 bdls. | The premium grade for roofs and sidewalls. 100% heartwood, 100% clear and 100% edge grain. |
| No. 2 Red Label | 16" (fivex) 18" (perfections) 24" (royals) | .40" .45" .50" | 20/20 18/18 13/14 | 4 bdls. 4 bdls. 4 bdls. | A good grade for many applications. Not less than 10" clear on 16" shingles, 11" clear on 18" shingles and 16" clear on 24" shingles. Flat grain and limited sapwood permitted in this grade. |
| No. 3 Black Label | 16" (fivex) 18" (perfextions) 24" (royals) | .40" .45" .50" | 20/20 18/18 13/14 | 4 bdls. 4 bdls. 4 bdls. | A utility grade for economy applications and secondary buildings. Not less than 6" clear on 16" and 18" shingles and 10" clear on 24" shingles. |
| No. 4 Undercoursing | 16" (fivex) 18" (perfections) | .40" .45" | 14/14 or 20/20 14/14 or 18/18 | 2 bdls. 2 bdls. 2 bdls. 2 bdls. | A utility grade for undercoursing on double-coursed sidewall applications or for interior accent walls. |
| No. 1 or No. 2 Rebutted-Rejointed | 16" (fivex) 18" (perfections) 24" (royals) | .40" .45" .50" | 33/33 28/28 13/14 | 1 carton 1 carton 4 bdls. | Same specifications as above for Number 1 and Number 2 grades but machine trimmed for exactly parallel edges with butts sawn at precise right angles. For sidewall application where tightly fitted joints are desired. Also available with smooth sanded face. |

**APPROXIMATE COVERAGE OF ONE SQUARE (4 BUNDLES) OF SHINGLES BASED ON FOLLOWING WEATHER EXPOSURES**

| LENGTH AND THICKNESS | 3½" | 4" | 4½" | 5" | 5½" | 6" | 6½" | 7" | 7½" | 8" | 8½" | 9" | 9½" | 10" | 10½" | 11" | 11½" | 12" | 12½" | 13" | 13½" | 14" | 14½" | 15" | 15½" | 16" |
|---|---|---|---|---|---|---|---|---|---|---|---|---|---|---|---|---|---|---|---|---|---|---|---|---|---|---|
| 16" x 5/2" | 70 | 80 | 90 | 100* | 110 | 120 | 130 | 140 | 150Z | 160 | 170 | 180 | 190 | 200 | 210 | 220 | 230 | 240† | -- | -- | -- | -- | -- | -- | -- | -- |
| 18" x 5/2¼" | -- | 72½ | 81½ | 90½ | 100* | 109 | 118 | 127 | 136 | 145½ | 154½Z | 163½ | 172½ | 181½ | 191 | 200 | 209 | 218 | 227 | 236 | 245½ | 154½† | -- | -- | -- | -- |
| 24" x 4/2" | -- | -- | -- | -- | -- | 80 | 86½ | 93 | 100* | 106½ | 113 | 120 | 126½ | 133 | 140 | 146½ | 153Z | 160 | 166½ | 173 | 180 | 186½ | 193 | 200 | 206½ | 213† |

NOTE: * Maximum exposure recommended for roofs.  Z Maximum exposure recommended for single-coursing on sidewalls.  † Maximum exposure recommended for double-coursing on sidewalls.
5/2" means that 5 shingles measures 2 inches at the butts.

### Red Cedar shingles
Table 11-9

## Valleys

Valleys are the most vulnerable area of shingle and shake roofs and should receive the most care and best materials. It is good practice to lay an 18 inch strip of 30 pound felt lengthwise in the valley before laying the metal. This will help prevent the metal from drawing condensation to the under side and rusting out before the roof wears out. Use 26 gauge or heavier galvanized metal that is preformed or W shaped. The metal should extend ten inches from the valley center on each side. On roofs with a pitch of 12 in 12 or more, the metal should extend at least 7 inches from the valley center. If more than one piece of valley material is used, provide at least a six inch lap at the joint. The valley shingles should have the grain running the same direction as the main roof shingles. Note Figure 11-10. You can save a lot of time by cutting all of your valley shingles at one time and stacking them on the roof. To keep the shingles straight in the valley, use a board laid along the center of the valley or snap a chalk line. A good valley is at least 5 inches wide and not over 6 inches wide at the top and bottom. Some recommendations are that the valley should be wider at the bottom than at the top. Most roofers agree that a valley equal in width from the top to the bottom looks best.

Driving staples with a gun
Figure 11-9

*Wood Shingles and Shakes*

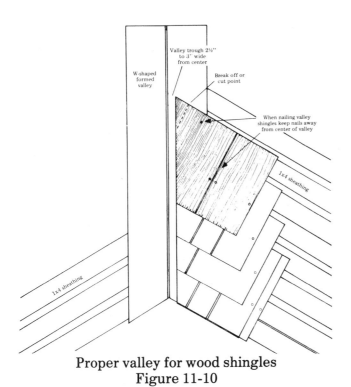

Proper valley for wood shingles
Figure 11-10

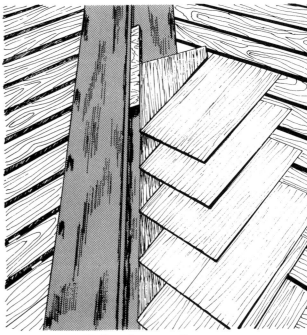

Valley for wet climates
Figure 11-11

Figure 11-11 shows a method used by some roofers in wet climates, especially where leaves and pine needles may build up in the valley. Roofers who recommend this method contend that it gives them a tighter and more waterproof valley. Note that the direction of the grain is running with the valley. Consequently this method should not be used in drier climates because the shingles will curl up and split.

Valley shingles should be selected from wide shingles and cut as neatly as possible. Some roofers cut their valley shingles one at a time with a knife or hatchet, but with a power saw you can cut two or more at a time. Determine the exact angle of cut by following Figure 11-12. The portion of shingle left over can usually be used on the hip ridge.

Trim the shingles back on the hips as you shingle if you are not using valley cutoffs. The hatchet makes a good cutting tool if it is kept sharp because you can use both hands to exert pressure. Note Figure 11-13.

### Shingling On Steep Roofs

When a roof is too steep to sit on safely, you may be able to get by with a job built roof seat, as in Figure 11-14. The seat illustrated was made from scrap lumber. The seat works best if its slope matches the roof slope exactly, giving you a level surface to sit on. Many roofers build only a seat for 12 in 12 roof slopes. Other roofers keep two or three seats with different angles in their truck. Make sure the seat you build and use is sturdy. Attach at least six beer can openers, point down, on the bottom side for good traction. The points will make small indentations in the shingles but will not harm the roof.

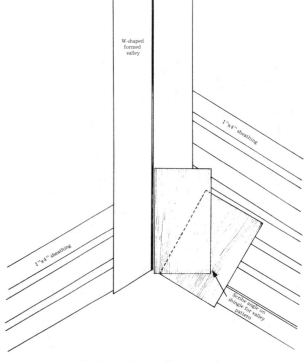

Determining valley angle
Figure 11-12

## Roofers Handbook

Trimming hip and ridge shingles
Figure 11-13

Roofer using a job built roof seat
Figure 11-14

The roofer in Figure 11-15 is shingling from a comfortable seated position. The belt around his left thigh is attached to the roof seat. This way he can move around easier. Note that he has a pile of shingles lying on the roof up against his roof seat. Whenever he needs more shingles, he has a pile on the sheathing immediately in front of him. The maximum roof pitch possible for using a roof seat depends on the roofer. Most roofers start scaffolding at about 12 inches per foot, but you may want to start at a lesser pitch. Do whatever works best for you on your particular roof, but do not overlook safety. The roofer here has a metal scaffold set up at the eave in case he should slip.

Use any of several methods to scaffold a steep roof. Usually you have to work with the materials you have. One of the more popular method uses wood shingles and 2x4 straight grain lumber about 10 to 12 feet long. For each board use three wood shingles about 5 inches wide, one on each end and one in the middle. Secure each shingle to the 2x4 with 6 wood shingle nails through the butt end. Then place the board on the roof where you need a scaffold and secure it as illustrated in Figure 11-16. This is normally a safe method but may depend on the roofer and the situation. If the roof is too steep, you may need to use a solid built roof jack, especially at the bottom. The scaffold made of wood shingles and 2x4 boards should be used primarily as a toe board and not to support great weight on a steep roof. Always be careful.

### Flashing Wood Shingles

Flashing must be made as permanent as possible because it should last as long as the roof and preferably longer. Even with the many modern sheet metal shops around the country, roofers still have to form their own metal in many cases. Figure 11-17 shows a typical dormer flashing job. The metal flashing at the bottom should be in one piece and step flashing should be used up the sides. The bottom piece

Comfortable shingling on a steep roof
Figure 11-15

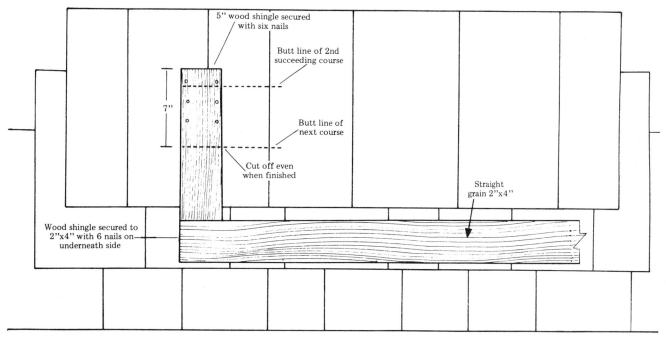

Toe board method on wood shingles
Figure 11-16

should extend onto the shingles about 5 inches. The amount turned up against the dormer will vary but should be at least 2 inches. The step flashing must be an individual piece for each course of shingles. It should be seven inches long for a 5 inch exposure and extend three inches onto the roof and up the wall. When placing step flashing, lay it under the butt of the shingle and nail the shingle in the normal areas. This way the nail should penetrate the top of the step flashing and hold it in place. Be sure that the nail does go through the metal. Always leave the metal at the bottom of the dormer uncovered and nail the bottom edge with as few nails as possible. Some roofers feel that the roof will look better if they lay another row of shingles on top of the metal to conceal it. Doing so increases the risk that moisture will collect under the shingles and rust the metal prematurely.

The bottom third of vent flashings should

Typical dormer flashing job
Figure 11-17

Shingles used to hold down flashing
Figure 11-18

Lead flashing on a soil pipe
Figure 11-19

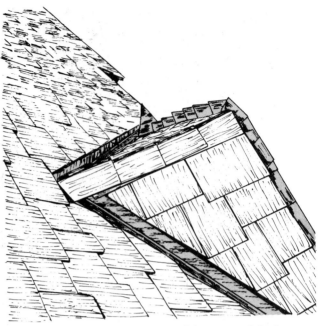

Proper flashing on a chimney cricket
Figure 11-20

also be left open. In Figure 11-18, however, the roofer has used shingles to hold down the bottom part of the flashing. This is equally acceptable roofing practice. Note, however, the one inch gap left around the pipe so that debris will wash free. Figure 11-19 shows another example of good craftsmanship, a lead flashing on a soil pipe. The roofer should go ahead and finish waterproofing the vent while he is in the area. Cut off the excess lead, leaving about ½ inch to be tapped down into the inside of the pipe.

Figure 11-20 illustrates a typical chimney cricket that has been shingled and flashed properly. The top of the valley has been ridged over to prevent water and snow from blowing under the shingles. Where the ridge stops at the main roof there is no way to get the joint tight enough to be watertight. Consequently the metal valley underneath must have one side lapped over onto the top of the other valley. Bend over enough flashing to have about a 4 inch lap and try to make this lap as neat and watertight as possible. If the joint is not watertight, slip a piece of tin up under the shingles and let it lap down enough to shed water. When you nail the last piece of ridge, place the two nails in each shingle through the butt portion.

### Ridge

The appearance of every roof depends on the neatness of the ridge. Moreover, a sloppy ridge usually means a bad roof. Factory ridge units make the job much easier because the two pieces are stapled together solidly and offset mitered joints are stacked alternately for easy application. Note Figure 11-21. Each piece of ridge or ridge unit must have two nails on each side, placed about seven inches above the butt edge. Give each ridge unit the same exposure as the roof shingles. Offset the ridge joints so that the ridge remains watertight if several joints

Factory ridge units
Figure 11-21

*Wood Shingles and Shakes*

Laying felt under ridge shingles
Figure 11-22

open up as the roof ages. The roofer in Figure 11-22 is taking the extra precaution of laying a narrow strip of 30 pound felt over the hip before applying the ridge shingles. Note that he is gauging the ridge and is also using a chalk line to keep the units straight.

Figure 11-23 shows a junction where two ridges meet. This should be handled as neatly as possible and with as few exposed nails as possible. When starting a new run, always start out with a double ridge, as in Figure 11-24. At junctions like this you may have to adjust the

Two ridges joining at a junction
Figure 11-23

Starting out at a double ridge
Figure 11-24

123

*Roofers Handbook*

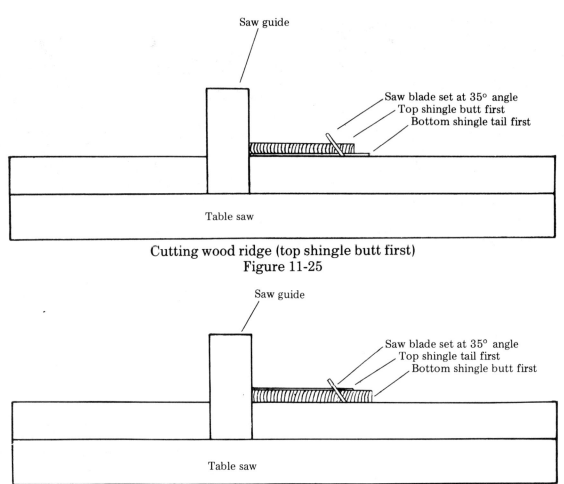

Cutting wood ridge (top shingle butt first)
Figure 11-25

Cutting wood ridge (top shingle tail first)
Figure 11-26

shingles somewhat to make a tight fit. Make the ridge meet in the middle and build a saddle by reversing two units on top of each other. Then trim back the tail ends and leave about 8 inches of the butt portion. Always use longer nails to apply the ridge so that the nails penetrate the sheathing.

Sometimes factory ridge units are not available. If the roof can't wait, you may have to cut your own ridge shingles. This takes some time and it is appropriate that the roofer charge extra for this work. Some builders will let you cut all the ridge units yourself if you don't charge extra. A table saw is needed and the blade should be set at about a 35 degree angle. In Figure 11-25 the saw guard was omitted so that the cut angle could be observed. Set the guide at 4½ inches or whatever width ridge you may need. Some roofers prefer about a 4 inch ridge. Cut two shingles at one time with one shingle butt first and the other tail first. The top piece of ridge will come out about ¼ to 3/8 inch wider than the bottom piece. The next piece will be reversed, as in Figure 11-26. This will give you alternating mitered joints like the factory made ridge. Cut as much ridge as you will need, alternating each cut every time. At 5 inch exposure you will need 24 units or pairs for every 10 feet. As you lay the two piece ridge, lay the narrow piece first and make it flush with the roof so that the other piece will lie flat. Use two nails per piece.

### Dutch Weave Pattern

This is sometimes referred to as a "thatch" or the "shake look" roof. Figure 11-27 shows a poor example of this type of roof. There is some consistency to the pattern in the middle but many shingles are protruding too far. In the fourth row from the top the shingles that are overhanging 7 or 8 inches will soon curl up and probably break off in the wind. The proper overhang is about ½ inch and never over 1 inch. A consistent pattern is the key to a very good looking Dutch weave roof. Generally you should nail on an extra shingle about every fourth shingle. The protruding shingle should fall directly between two protruding shingles in the

Poor example of "thatch" or "shake look" roof
Figure 11-27

Using a notch on hatchet handle to gauge felt
Figure 11-28

course below. Don't leave large spaces without overhanging shingles as did the roofer here. This is very noticeable.

### Applying Shakes

Proper shake application is easy though many roofers have trouble making a neat and watertight roof. The application methods are basically the same as with wood shingles. However, with shakes at standard double coverage exposures you must apply an 18 inch strip of 30 pound felt, overlapping the tops of each course of shakes 4 inches. This applies both to roofs with solid sheathing and spaced sheathing. Solid sheathing should be used in areas that have blowing snow. A complete felt underlayment on low pitch solid deck roofs may be beneficial. In any case a 36 inch strip of 30 pound felt must be used at the eaves. Keep the 36 inch strip flush with the very bottom edge for complete protection and use only enough nails to keep it in place. Use large head roofing nails (7/16") for nailing the felt. The first strip of 18 inch felt must be positioned so the bottom edge is 20 inches above the butt edge of the first row of shakes. If the 36 inch felt was applied with the bottom edge flush with the edge of the eave, no measurement will be necessary. The top edge of the half strip will fall on the top edge of the 36 inch strip. Keep the half strip straight as this is your guide for laying the shingles. Apply felt to as much area as you plan to cover that day and only nail the felt at the top. If wind is a problem you can nail down band sticks or narrow shakes to hold the felt in place. Place one nail in each stick or shake and make sure that the nail is 1 inch above the bottom edge of a lap. Later, when you shingle up to the stick, pull the nail and place a shingle under the lap and squarely over the hole. This will be watertight because there will be two more layers of shakes over the hole. It is better, however, if bundles of shakes are used to hold the felt in place.

Lacing felt on hips
Figure 11-29

Turning back the ends on the half strips
Figure 11-30

In Figure 11-28 the roofer is using a 10 inch notch on his hatchet handle as a guide to lay the felt. This is good enough as the felt need not be perfectly straight. If the felt gets out of line a little, as it may on a wide slope, measure up from the eave to get straightened out. About half way between the ridge and eave you should stop and measure from the bottom edge of the last strip to the top of the ridge. Adjust the remaining strips so that the top of the last strip will extend over the ridge about 3 inches. This way the next to last row of shakes will just come to the ridge. Then you can finish off with 15'' starter-finish shakes. After the ridge is applied this will leave a ten inch course at the top. This also saves a lot of unnecessary cutting. If 15'' starter-finish shakes are not available in your area, cut what you need on the ground.

For extra protection, the hips should be laced together as in Figure 11-29. If the felt is watertight, the roof should be watertight. There is virtually no way to be too thorough when applying a shake roof, especially one that is expected to last for several decades. Usually when shake roofs fail it is not because the shingles are worn out. Instead either the felt or the metal flashings have failed.

Figure 11-30 shows a trick that is used by experienced shake roofers when roofing over spaced sheathing. Doubling back the ends on the half strip makes an extra layer which prevents the felt from sagging into the spaces between the sheathing. Also, the felt doubled back will repel any water that might get that far under the shingles. It is a simple task and may prevent a costly leak. The felts turned back around this large vent are the felts that will remain under the vent flashing when the shingles are applied.

Any necessary laps in the open roof should also be doubled back. Note Figure 11-31. If you allow a 12 inch lap you should never have any trouble with leaks, at least at the lap. Doubling back the ends of the half strip is a good

## Wood Shingles and Shakes

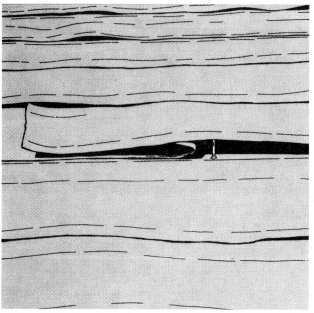

Doubling back unnecessary laps in open roof
Figure 11-31

precaution and can prevent a leak in certain circumstances. Many leaks are of an unusual nature and sometimes unusual methods are needed to prevent these leaks.

For extra protection in snow troubled areas, many roofers use an extra half strip at the bottom, as in Figure 11-32. This allows any moisture that gets under the shingles to run out on top of the starter shingles. Many roofers make it a habit to use this method on every roof. This extra half strip provides extra protection and serves as a guide for the second row of shingles. When this method is used it is easier to first apply the 36 inch strip of 30 pound felt to the eave and then apply the starter shingles before laying the half strip. This first half strip is positioned 8 inches up from the eave or 10 inches from the butt edge of the starter shingles if a 10 inch exposure is required. Note that the extra half strip will cover the nails in the starter shingles.

When applying the first row of shakes on top of the starter shakes, select only shakes that will permit you to offset the joints of the starter at least 1½ inches. Maintain this margin on the entire roof. Place your nails about 2 inches above the projected butt line of the following row of shingles and 1 inch in from each edge. Always use only two nails per shingle regardless of the width. Work above the shingles from off

### CERTI-SPLIT RED CEDAR HANDSPLIT SHAKES

| Grade | Length & Thickness | 18" pack** | | Description |
|---|---|---|---|---|
| | | Number courses per bundle | Number bundles per square | |
| No. 1 Handsplit & Resawn | 15" starter-finish | 9/9 | 5 | These shakes have split faces and sawn backs. Cedar logs are first cut into desired lengths. Blanks or boards of proper thickness are split and then run diagonally through a bandsaw to produce two tapered shakes from each blank. |
| | 18" x ½" to ¾" | 9/9 | 5 | |
| | 18" x ¾" to 1¼" | 9/9 | 5 | |
| | 24" x 3/8" | 9/9 | 5 | |
| | 24" x ½" to ¾" | 9/9 | 5 | |
| | 24" x ¾" to 1¼" | 9/9 | 5 | |
| No. 1 Tapersplit | 24" x ½" to 5/8" | 9/9 | 5 | Produced largely by hand, using a sharp-bladed steel froe and a wooden mallet. The natural shingle-like taper is achieved by reversing the block, end for end, with each split. |
| | | 20" pack | | |
| No. 1 Straight-split | 18" x 3/8" True-edge* | 14 straight | 4 | Produced in the same manner as tapersplit shakes except that by splitting from the same end of the block, the shakes acquire the same thickness throughout. |
| | 18" x 3/8" | 19 straight | 5 | |
| | 24" x 3/8" | 16 straight | 5 | |

Note: * Exclusively sidewall product, with parallel edges. ** Pack used for majority of shakes.

APPROXIMATE SQUARE FOOT COVERAGE OF ONE SQUARE BASED ON THESE WEATHER EXPOSURES:

| Shake type, length & thickness | 5½" | 6½" | 7" | 7½" | 8½" | 10" | 11½" | 14" | 16" |
|---|---|---|---|---|---|---|---|---|---|
| 18" x ½" to ¾" handsplit & resawn | 55(a) | 65 | 70 | 75(b) | 85(c) | 100(d) | -- | -- | -- |
| 18" x ¾" to 1¼" handsplit & resawn | 55(a) | 65 | 70 | 75(b) | 85(c) | 100(d) | -- | -- | -- |
| 24" x 3/8" handsplit | -- | 65 | 70 | 75(e) | 85 | 100(f) | 115(g) | -- | -- |
| 24" x ½" to ¾" handsplit & resawn | -- | 65 | 70 | 75(a) | 85 | 100(h) | 115(g) | -- | -- |
| 24" x ¾" to 1¼" handsplit & resawn | -- | 65 | 70 | 75(a) | 85 | 100(h) | 115(g) | -- | -- |
| 24" x ½" to 5/8" tapersplit | -- | 65 | 70 | 75(a) | 85 | 100(h) | 115(g) | -- | -- |
| 18" x 3/8" true-edge straight-split | -- | -- | -- | -- | -- | -- | -- | 100 | 112(i) |
| 18" x 3/8" straight-split | 65(a) | 75 | 80 | 90 | 100(g) | -- | -- | -- | -- |
| 24" x 3/8" straight-split | -- | 65 | 70 | 75(a) | 85 | 100 | 115(g) | -- | -- |
| 15" starter-finish shakes | Use supplementary with shakes applied not over 10" weather exposure. | | | | | | | | |

(a) Maximum recommended exposure for 3-ply roof construction.
(b) Maximum recommended exposure for 2-ply roof construction; 7 bundles per sq. at 7½" exposure.
(c) Maximum recommended exposure for sidewall construction; 6 bundles per sq. at 8½" exposure.
(d) Maximum recommended exposure for starter-finish course application; 5 bundles per sq. at 10" exposure.
(e) Maximum recommended exposure for roof pitches between 4 in 12 and 8 in 12.
(f) Maximum recommended exposure for roofs steeper than 8 in 12.
(g) Maximum recommended exposure for single coursed sidewall construction.
(h) Maximum recommended exposure for 2-ply roof construction.
(i) Maximum recommended exposure for double coursed sidewall construction.

Table 11-31

Extra half strip at bottom
Figure 11-32

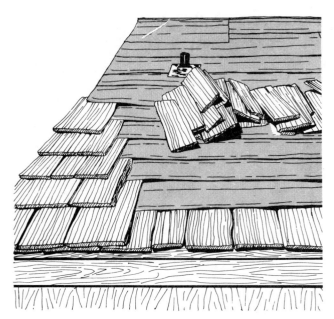

Applying wood shingles
Figure 11-33

the felt to apply just enough shingles so that you can then sit down to work comfortably. Note Figure 11-33. As you shingle, try to determine exactly how many courses you can reach with ease. Applying too many courses at once will slow you down. Taking only 1 or 2 rows at a time will also be slow. Before sitting down to shingle, make sure that you have plenty of shingles and felt laid out in front of you.

Figure 11-34 shows a solid sheathed roof with the application of 30 pound felt underlayment and shakes just started. If you have a helper working with you, get him started laying felt and then begin shingling while he continues felting in the roof slope. As soon as your helper lays enough felt he can drop back and distribute some shingles for you before laying more felt.

In Figure 11-35 the roofer is working alone and has felted one half of his slope before starting to apply shakes. The slope is just steep enough to make loose shingles slide. This will slow the roofer down some. In this case a toe board at the bottom edge would help keep the shingles from sliding off the roof even though the roofer may not need it himself.

### Valleys

You can save a lot of time by precutting the valley shingles and stacking them on the roof for future use. Cut the shingles from wide choice shingles. The best method is to determine the angle as was done in Figure 11-12, scratch the shingle with a nail or knife and cut with a power saw. The roofer in Figure 11-36 has laid a half strip of 30 pound felt lengthwise in the valley prior to the application of the half strips. This lengthwise strip serves as a guide for laying the valley shingles. The roofer in Figure 11-36 allowed a three inch margin on each side of the center. Normally you should allow about four inches but the 3 inch margin still makes a neat and efficient valley, as is evident in Figure 11-37. To help prevent moisture from condensing on the valley metal and rusting it out eventually, a layer of 30 pound felt should be

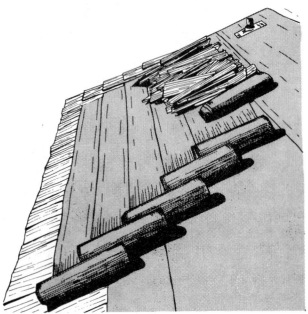

A solid sheathed roof with felt and shakes started
Figure 11-34

Working a steep roof
Figure 11-35

laid on top of the roof sheathing prior to installing the valley.

Flashing

Many shingle roofs fail because the wrong kind of metal flashing is used or the right flashing is installed wrong. The current trend is to use continuous wall flashing because it is easy to apply and the builder doesn't have to wait on the roofer. The builder can apply his siding and the painter can do his painting before roofing is applied. This continuous wall flashing does look

Laying valley shingles
Figure 11-36

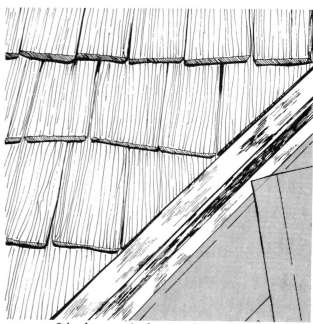
3 inch margin from valley center
Figure 11-37

Improper flashing
Figure 11-38

A leak prone situation
Figure 11-39

good but is prone to develop very serious leaks even if special precautions are taken. Where wall flashing is needed, all shingles must have individual step flashing. Many builders get by temporarily using continuous flashing on composition roofs and some builders assume that this is adequate for wood roofs also. Some roofers make the flashing waterproof for a short time by sealing the shingles with plastic cement. This cement will not adhere for long on wood shingles or shakes. The wood soaks up water, expands, and releases the plastic cement. Many builders are not aware of what will happen and should be advised of the need for step flashing. The flashing in Figure 11-38 will leak because any water that reaches the metal has no opening to escape. The water will carry dirt and debris under the shingles. Eventually this dirt and debris will form a dam and back up the water until it leaks through the felt and into the house. The dirt will retain moisture which will rust the metal, eventually causing a more serious leak. Individual step flashing will turn the water out onto the shingle below, therefor eliminating leaks. If you encounter a situation such as this, individual step flashing can still be installed by loosening the siding and slipping the metal up under it. Leave the continuous flashing in place and simply cover it with new step flashing at least two inches longer than the exposure. Make sure the step flashing goes six inches onto the shingle and about five inches up the wall. The wall coverage figure may be reduced when you are trying to work it under siding, but never use less than two inches. When you cover up continuous step flashing, take your hammer and flatten down the water guard so the shingles will lie flat.

To prevent being called back for a leak unnecessarily, always watch for potential leaks others may have created. Figure 11-39 shows a sure leak. It will appear to be the roofer's and he will be called back to investigate. The roofer would then have to convince the builder or homeowner that the problem is the siding, not the roof. It would be easier in the beginning to inform the builder or homeowner of the potential problem. This siding will leak because it was not lapped enough to allow for shrinkage and cracked because it was improperly nailed.

You can be sure that the flashing will last as long as the wood roof you are applying if you use either heavy galvanized metal or copper flashing. Galvanized metal should be prepainted on both sides, using a good grade of metal primer. Clean the metal thoroughly with mineral spirits before applying the primer. Sharp bends should be made prior to painting to prevent the paint from cracking when the metal is bent.

### Vents

Most leaks occur around vents in about every type of roof, usually because of inadequate shingling around the vent. The standard vent flashing that is usually supplied to the roofer is actually not as wide as it should be. In Figure 11-40 the roofer laid a medium width shingle on both sides of the vent pipe, notching each

Vent pipe surrounded by medium
width shingles
Figure 11-40

Double layer of felt applied to flashing
Figure 11-41

shingle just enough to fit around the pipe. Then the vent flashing was installed and two nails were placed at the top to hold it in place. In Figure 11-41 a precaution was taken that usually is worth the time. A double layer of 30 pound felt was cut, slipped under the half strip and lapped over the top of the flashing. Now, if the shingles around the vent should crack, the vent still won't leak.

In Figure 11-42 the shingles have been applied properly. At the top of the vent a wide shingle was used to cover the flashing. This shingle was lowered below it's normal position so that it lapped over the top edge. Under the wide shingle a 15" starter shingle was used to fill in. A one inch gap was left around the vent flashing so that dirt and debris would wash free. The lowered shingle will not be noticeable from the ground.

### Hips And Ridges

The usual method of shingling on hip roofs is to bring the shingle all the way to the point of the ridge so you have something to nail to. Often the pointed part of the shake will split out and not be of much help. Figure 11-43 shows a method used by some shake roofers. They hold the shakes back from the edge slightly, permitting the ridge to lie a little lower. This way the nails penetrate the sheathing further, making a tighter ridge. Always use longer galvanized nails for nailing the ridge. Usually 8 penny nails are enough.

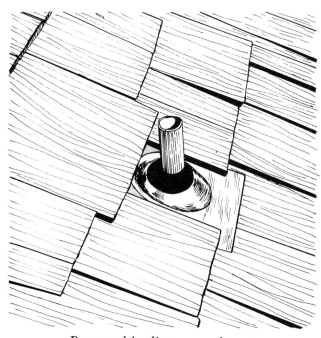

Proper shingling around vent
Figure 11-42

To apply the ridge material to the hips, start out at the bottom with a double layer of shake ridge units. The starter or first piece of ridge should be cut back so that it does not lap over onto the second row of shingles. This will allow you to lay a smoother, straighter ridge. See Figure 11-44. Place two nails in each shingle so that they will be covered at least two inches by the next piece of ridge. To keep the ridge fairly straight, use a chalk line on one side of the ridge. Factory ridge is easier to lay as each unit

Method of nailing ridge
Figure 11-43

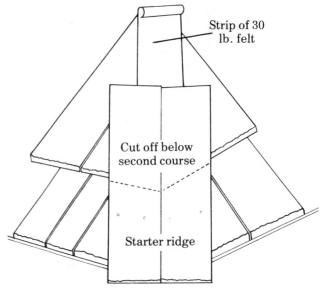

Starting the hip ridge
Figure 11-44

Hip ridge as viewed from ground
Figure 11-45

is stapled together and made from more uniform shingles. If factory ridge is not available, select the best shingles and split pieces 5 inches wide with your hatchet. Prior to laying the ridge put down a six inch strip of 30 pound felt to prevent a possible leak if the ridge should split or gap open.

Use a chalk line as you lay the hip ridge, but watch the other side of the ridge shingles for any unit that is wider or narrower than normal. Narrow hip ridge shingles can be set aside and used at the top ridge where they won't be noticed. The hip ridge will be very evident from the ground. Note Figure 11-45. Wide ridge shingles can be narrowed by splitting off the excess with your hatchet. If the ridge must be perfectly straight, carefully tack a 1x4 board in place on each side and build the ridge to fit between these boards. This will not occur too often because usually a shake roof is selected because of its rustic look rather than its neat, precise lines.

Figure 11-46 shows a task that most roofers can avoid. Usually 18 inch half strip felt will be available at your lumber yard. Some builders try to get by without spending the small premium for half strip felt. The roofer here should charge extra for cutting the felt. Cutting consumes time and requires a good power saw with a carbide tipped blade, the type used for masonry.

Shake roofs are desired because of their overall rustic appearance. However, neatness and straight courses are necessary to avoid a cobbled up look, particularly on steep roofs. Note Figure 11-47. Before you start a roof like this, look around the building and do some planning. Think of the shakes as though they were siding. The courses must wrap around the building as if it were a hip roof. Keep the courses even, regardless what type of ridge is used. If no ridge is used, use the woven or mitered corners illustrated in Figure 11-48. This is possible only if the roof is steep enough to prevent leaks. If in doubt, use an extra layer of 30 pound felt under each joint or apply galvanized metal flashing under the corners.

On side walls there are two appropriate inside corner methods, the jointed inside corner and the woven inside corner. Note Figure 11-49. You can use these methods on either wood

## Wood Shingles and Shakes

Cutting felt
Figure 11-46

A steep roof
Figure 11-47

Woven and mitered corners
Figure 11-48

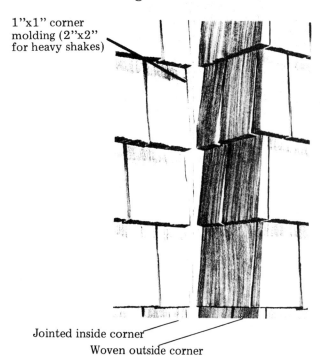

Jointed and woven corners
Figure 11-49

shingles or shakes. For shake wall application, the maximum recommended weather exposure with single coursed wall construction is 8½ inches for 18 inch shakes and 11½ inches for 24 inch shakes. Shakes can also be applied double-coursed, with an underlay of regular cedar shingles for each course. With such construction, 18 inch shakes can be applied at weather exposures up to 14 inches, and 24 inch straight-split shakes can be exposed up to 20 inches, but use a 24 inch cedar shingle underlay in each case. Butt-nailing is necessary for double-coursed application. Concealed nailing is customary with single-coursed construction. In either case, do not drive the nail heads into the surface of the shakes.

# Chapter 12

# Operating a Roofing Company

The roofing business is relatively easy to get into in that a minimum investment is required. Many roofers begin with little more than a truck, some tools, a telephone and an answering service. Overhead can be kept to a minimum if you work out of your home and your truck. You hire employees and buy materials only as required for the job you are working on. Your wife can keep your books for you though you may need an accountant to set up your records and prepare your tax returns. Don't be reluctant to continue to work as a roofer for some other contractor until you have developed enough work of your own to keep you busy. Once you have a growing, successful roofing business, you have the independence and security that employees seldom enjoy. Not everyone can make it, but thousands do every year. There are few businesses which consistently yield so large a profit on so small an investment.

## Estimating

You won't succeed as a roofing contractor if you can't estimate the labor and material quantities required. There are many "estimating systems" for roofing, but most roofers agree that certain principles apply no matter how you compile the estimate. On reroofs, all measurements should be made while on the roof. This is important as the estimator or roofer must determine the condition of the old roof and how many layers of shingles are on the existing roof. Try to spot any framing, sheathing or fascia that need to be replaced prior to roofing. Also, check the condition of the vent caps, vents and flashings. All these must be considered when figuring a reroof and should be included in the estimate. Usually the homeowner appreciates being informed of the condition of the chimney, the vents and the soundness of the structure overall. This tells him that you have been thorough in your investigation and that you are a professional. The homeowner would usually prefer that the roofer replace the vent caps as well as the flashings when needed. If the fireplace or chimney is very old, it may need some repair such as "tuck pointing" of the mortar joints, a new seal cap on top, or a flue cap to prevent rain from entering the flue tile. Since any water that gets into the house is as-

## RECOMMENDED REROOFING GUIDELINE

| The New shingles | The Existing Shingles Prior To Reroofing | | | |
|---|---|---|---|---|
| | Wood | 3-tabs & strip shingles with no C/O | Lock shingles | Dutch lap & hexagon shingles |
| Wood | In all areas except in wet climates. Then the old shingles should be removed. | Yes | Yes | Yes but a tear off would make a smoother roof. |
| 3-tabs and strip shingles. | Yes if the butt-up method can be used. If not then the old shingles should be removed. | Yes if the butt-up method can be used. If not then the old shingles should be removed. | No. Tear off the shingles or use another shingle. | No. Tear off the shingles. |
| T-locks | No. Use a 3-tab, strip shingle or a wood shingle. | Yes | Yes. But offset the headlaps. | Not desirable. A tear off would make a smoother job. |
| Shakes | Yes | Yes | Yes | Yes |
| The new asphalt strip shingles (shake appearance). | If the butt-up method can be used. Some brands have a different weather exposure and the butt-up method can not be used then. | Yes | Only the heavier shingles should be used. | Only the heavier shingles should be used. |

Note: These recommendations are based on experience.

Figure 12-1

sumed to be the fault of the roofer, examine the entire roof area carefully and advise the homeowner accordingly. The last step before writing up the estimate is to carefully walk around the house looking for any wood that should be replaced or for a small porch roof that you didn't notice from on the roof. Notice anything else that could cause your crew to be delayed until repairs are made. Noticing a problem in advance can turn a possible loss of time and money into a small profit.

### Roof Covers

Normally three shingle roofs are the maximum allowed on any structure. This rule is followed by many city regulations and most insurance companies. However, some houses should never have more than two roofs and there are some homes that support four roofs very nicely. Some houses are not strong enough to carry the weight of two roofs. The reroofing capacity depends on the soundness of the structure and the shingles involved. For instance, a house with two wood shingle roofs will not have as much weight on the rafters as another house the same size with one asphalt shingle roof. This is because each wood roof weighs only about 145 pounds per square (16" shingles). Also, the roof sheathing is probably spaced and creates a smaller load per square. Examine the roof carefully and determine if the rafters are sagging. Some bracing may have to be done prior to reroofing. Take every precaution as the homeowner is relying on your good judgment.

If the structure is solid and the house will easily hold the second or third roof, then make a determination of what shingle will be most compatible with the shingles on the roof. Actually, you can apply any shingle over a wood or asphalt shingle but the finished roof will not be sound with some combinations. For example, 3-tabs will be uneven if laid over T-locks or hexagon shingles, and T-locks will sag between the butt edges of a wood roof. Using the wrong shingle will probably reduce the expected life of the new roof. If the homeowner is determined that you use the wrong shingle, advise him

*Roofers Handbook*

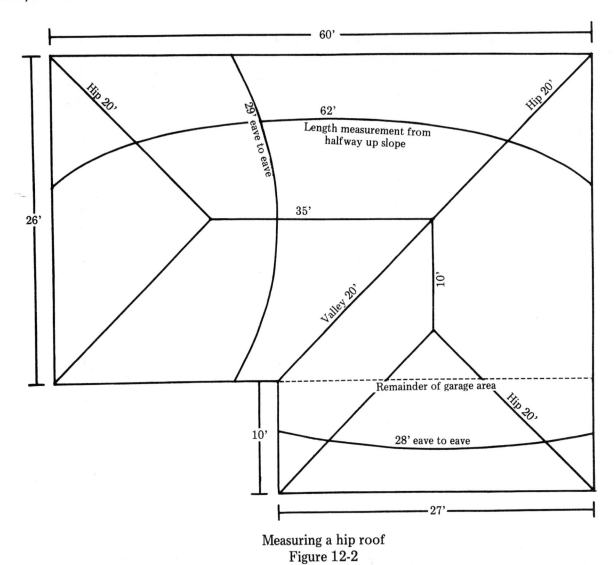

Measuring a hip roof
Figure 12-2

about what he can expect. Actually, you may be better off to refuse a job under these circumstances. Putting on a bad roof won't please a homeowner no matter what he insisted on earlier. The neighbors will spread the word about the poor roof and your reputation is bound to suffer in spite of the true facts. When planning a reroof, use the reroofing chart in Figure 12-1. In the old roof category, tile, shakes, asbestos and slate are not included because you cannot reroof over any of these surfaces. You must tear off the old roof before reroofing. When reroofing with slate, tile or asbestos you should consult the manufacturers for specific details on the job you are planning.

### Hip Roofs

To estimate a hip roof, you need not measure each slope individually. You can arrive at a good overall estimate that is quite accurate very easily. Measure the length laying the tape half way up the hip and measure the width from one eave up and over to the other eave. See Figure 12-2. This measurement will allow for some of the material that will be wasted but the estimator should figure one square extra to be sure. This method will be accurate for most houses. Allow extra shingles for irregularly shaped houses or cut-up roofs. The hip roof in Figure 12-2 measures 60 feet long at the eave. Laying the tape half way up the hip the length measurement is 62 feet. The house is 26 feet wide and the width measurement up and over the ridge is 29 feet. These two figures omit a portion of the garage which should be measured the same way: 10 feet by 28 feet. Now you can figure the square foot area by multiplying the width of the main roof times the length. Then multiply the width of the garage times the length. Total the two figures. This sum is the total square foot area of roof. To find the squares of shingles needed divide by 100 or move the decimal two figures to the left.

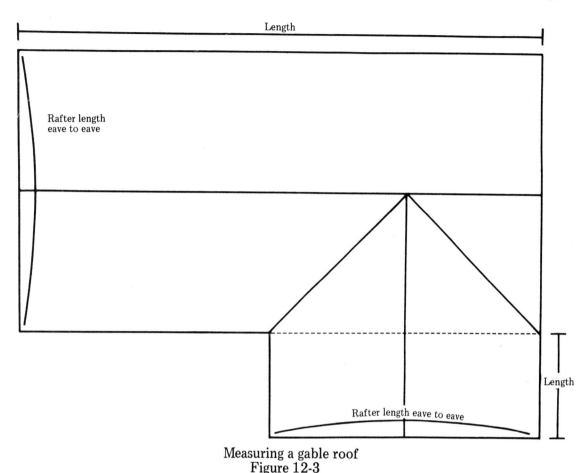

Measuring a gable roof
Figure 12-3

| Main roof | Garage area | Total of the two areas |
|---|---|---|
| 62 ft. | 28 ft. | 1798 sq. ft. |
| x29 ft. | x10 ft. | +280 sq. ft. |
| 1798 sq. ft. | 280 sq. ft. | 20.78 squares |

Insert decimal

Shingles are sold by the square and one square covers 100 square feet. The 20.78 squares should be rounded up to 21 squares. The waste and the ridge and starter material needed are not yet included in the total figure.

As you measure the roof, make a drawing of the entire roof area, detailing the valleys, vents, chimney and anything else that should be remembered. On the sketch you should show the lengths of the ridges and valleys so you can complete the estimate. Make complete notes. Quite often an estimate will be filed for several months before the homeowner decides to reroof his house.

Now figure the waste, ridge and starter material. For the roof in Figure 12-2 the waste will be about one square which is the average waste for a normal hip roof. For the starter course you can figure a bundle of 3-tab units for every 80 lineal feet. A starter roll 9 inches wide, with 4 units to a 36 inch roll (one square) will cover 144 lineal feet, minus the laps. A bundle of 3-tab units will ridge 33 to 35 lineal feet. You can figure one square of 3-tabs or standard hip and ridge units for every 100 lineal feet.

Totaling the eave measurements in Figure 12-2, 190 lineal feet of starter course will require one square of 3-tabs. The ridge totals 145 lineal feet and will require 1 2/3 squares of 3-tabs. The material estimate can now be completed as follows:

21 sq. shingles (roof area)
 1 sq. shingles (starter course)
1 2/3 sq. shingles (ridge)
 1 sq. shingles (waste for average house)
24 2/3 sq. shingles

This 24 2/3 squares is the amount needed to roof the house. Many roofing contractors now multiply this figure by a base price that they need to receive to operate successfully. This base price per square includes all costs: labor cost per square, cost of shingles per square, nail cost per square (approximately 2 pounds per square), cost of delivery and clean up, sales commission if any and the necessary profit and overhead. This base price applies to the average house under ordinary circumstances. Any additional material or labor costs such as for valleys, metal edging, tear offs, steeper slopes, the replacing of any vents or wood, or the patching in of missing shingles should be added to the estimate separately. A well detailed estimate will show the customer where his money is going and will make the sale much easier.

To measure a gable roof such as in Figure 12-3, multiply the roof length times the rafter length. The amount of waste on a gable roof can vary greatly depending on how the shingles end up at the finishing rake, the type of shingle used, the pattern used and of course the roofer. Some roofers on certain jobs will have almost no scrap left over; others may waste two or three squares of shingles on identical houses. When measuring, you should round your measurements off to the next higher even foot. Then you need to include only about 5 percent for waste on the average house and 10 percent on a cut-up house. If you pay your roofers by the square, inform them in advance that you are paying them for the estimated area, not the squares of shingles they use. This way they won't mind having some shingles left over and will usually try harder to not run short. A working arrangement such as this has been used by many roofing contractors with good success.

The amount of waste that you should figure for valleys depends on the type of valley used. An open valley will require almost no additional material. A closed valley (half or full lace) will require about one bundle for every 25 lineal feet.

While it is advisable to measure each roof from the roof surface, it is sometimes impossible to do so because the roof is too steep. You can always measure the length from the ground. But to measure the rafter length you will have to place a ladder at the eave and extend a tape measure up the roof. If the tape measure is not long enough, you can count the remaining shingles to determine the complete length of the rafter. Be sure of the exact exposure of the shingles when estimating a slope like this.

When estimating new roofs, use the same methods as described for reroofs. But get the measurements from the plans. Look at the end view to get the rafter length and the front or rear view to obtain the length of the roof slope. Use a scale ruler matched to the scale of the plans. Look at every angle of the structure and make a sketch showing all details as though it were a reroof job. If only a floor plan is available, leave the total number of squares open until the job is finished.

With styles in architecture becoming more varied, you may need to estimate a triangular roof. Any triangle is one-half of a parallelogram which has the same length and height. Note Figure 12-4. Simply measure the length and width (or height), multiply the two figures and then divide by 2. The answer is the roof area.

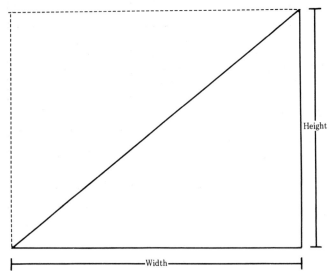

Estimating a triangular roof
Figure 12-4

Labor Costs

Some roofers work for a certain wage per square while others work by the hour. To estimate the hours per square, you must consider many variables: the shingle type used, the pitch of the roof, the overall height of the structure and the general style of the structure. After all of this is considered, you will still have variable factors such as the experience of the applicators and the possibility of weather problems. Table 12-5 gives manhours per square and is based on a typical roofer's work on an average house.

By maintaining records you can easily develop productivity figures for your applicators. The average roofer should be able to lay shingles at the rates listed in Figure 12-5. However, a shingling specialist will exceed these figures. Applicators who lay shingles every day will average between one and two squares per hour and sometimes more. If your applicators can not meet the times in Table 12-5, you may need to hold a training session.

If you pay by the hour, remember the "labor burden" that you as an employer must carry for each hour of wages paid. The "labor burden" will add between 25% and 30% to the labor

## Operating A Roofing Company

| Type Of Roofing | Average Roof 1 Square | Cut Up Or Difficult Roofs 1 Square |
|---|---|---|
| Asphalt strip shingles | 1-2 hours | 2-3 hours |
| Lock type asphalt shingles | 1-2 hours | 2-3 hours |
| Wood shingles | 2 hours | 3-4 hours |
| Shake shingles | 1-2 hours | 2-3 hours |
| Slate shingles | 3-5 hours | 5-8 hours |
| Spanish or Mission tile | 6-8 hours | 8-10 hours |
| Asbestos rectangular shingles | 2-4 hours | 4-6 hours |
| Aluminum shingles | 1-2 hours | 2-3 hours |

Hours per square
Table 12-5

cost. For every dollar of payroll you must pay an additional 25 to 30 cents in taxes and insurance to government agencies and insurance carriers. Many roofers add 30% to the estimated labor cost to cover taxes and insurance. Most states levy an unemployment insurance tax on employers based on the total payroll for each calender quarter. The actual tax percentage is usually based on the employer's history of unemployment claims and may vary from less than 1% of payroll to 4% or more. The Federal government also levies an unemployment insurance tax based on payroll (F.U.T.A.). The tax has been about .7% of payroll. The Federal government also collects Social Security (F.I.C.A.) and Medicare taxes. Together these come to about 6% of payroll depending on the earnings of each employee and are collected from the employer each calender quarter or more frequently.

States generally require employers to maintain Worker's Compensation Insurance to cover their employees in the event of job related injury. Heavy penalties are imposed on employers who fail to provide the required coverage. The cost of the insurance is taken as a percentage of payroll and is based on the type of work each employee performs. Clerical and office workers have a very low rate classification and the employer's cost may be only a small fraction of 1% of payroll. Hazardous occupations such as roofing carry a rate of about 20% of payroll. Most light construction trades have a rate between 4% and 8%. The actual cost of Worker's Compensation Insurance varies from one area to another and from one year to the next depending on the history of injuries for the previous period. Your insurance carrier will be able to give you the cost of coverage for the type of work your employees are doing.

Every contractor should maintain liability insurance to protect his business in the event of an accident. Liability insurance is also based on the total payroll and usually is about 2.0% of payroll. Higher liability limits will cost more.

The total "labor burden" can be itemized as follows: (The percentages listed are approximate maximums. Your accountant or bookkeeper will have more exact figures.)

State Unemployment Insurance .........4.0%
F.I.C.A. and Medicare .................6.1%
F.U.T.A. ............................0.7%
Worker's Compensation Insurance .....22.0%
Liability Insurance ..................2.0%
Total contractor burden .........34.8%

Every contractor who has a payroll is well advised to add this "labor burden" into the cost of labor on every estimate and make the insurance and tax deposits when due. No contractor can ignore these requirements and operate for long.

### Direct Overhead

Regardless of how your crews are paid, your cost of doing business is far more than the cost of labor, materials and equipment. There are many other costs which you must bear that are incurred as a result of taking each particular job. These costs are usually called *direct overhead* and can be thought of as administrative costs. The list below includes some of the items that are usually included as direct overhead.

- Roof Bonds
- Pensions
- Building Permits
- Dump Charges
- Sales Commissions
- Supervision
- Repairs to adjoining property
- Job toilets

You can probably think of many more direct overhead items. Some roofers include in direct overhead the cost of supervision and other nonproductive labor such as the cost of estimating the job. The time you spend on each job should be charged against each job. These are very real costs and must be included somewhere in the estimate. Since they are incurred as a result of taking each particular job, they can be properly included under direct overhead.

### Indirect Overhead

After everything is figured, there are certain items of expense you must bear in conducting your business which cannot be charged directly against any certain job. For example, office rent, telephone at the office, office lighting, office staff, small tools, office insurance, printing, service to vehicles, postage, telegrams, and countless other items. Some roofers favor taking the total of each month of such unassigned expense and assume it as the cost of doing a certain amount of work. Then they reduce this figure to a daily expense. If you are doing two jobs of about the same size on a given day, each would bear one-half the indirect overhead cost for that day. Some roofers figure the indirect overhead cost as a cost per productive manhour or a percentage of the total job cost. Other contractors have reduced the indirect overhead to a cost per square of roof area. Any one of these systems is good if it works for you. Keep a record of your indirect overhead and develop some method of dividing this cost among your jobs. Most of all, don't forget to include this important item in your bid. The total cost of direct and indirect overhead will be 10% or more of the total job cost for most work. This 10% will make the difference between a profit and a loss on almost every job you have.

### Profit

The profit is the return on investment to the owner of the roofing company. The roofer should be able to pay himself a wage for the work he performs and, in addition, pay himself a return on the money he has invested in his business. If you have $50,000 invested in your business you should receive a return on investment of $3,200 to $6,000 per year (8% to 12% of investment) in addition to a reasonable wage. This profit can be thought of as interest on the money invested in equipment, office, inventory, work in progress and everything else associated with running a roofing business. How much then, should you include in your estimate for profit? You will hear many conflicting figures. Some estimators try to end up with a "profit" of 20 percent of the total job price after all bills are paid. They may operate efficiently enough to achieve a 20 percent profit. They certainly are the exception. The roofing contractor who talks about a 20 percent profit may mean that after he has paid for his labor, material and equipment he has 20% left over for himself. This 20% is really his wage and, though it may be substantial, it is not a profit in the true sense. A profit is what remains after *all* costs are considered. The roofer should include the cost of his own work under *direct overhead*. What then is a realistic profit in the true sense? Dun and Bradstreet, the national credit reporting organization, has compiled figures on small contractors for many years. They report the average net profit after taxes for all contractors sampled to be consistently between 1.2% and 1.5% of gross receipts. This includes many contractors who reported losses or became insolvent. A 1½% profit, even after taxes, is a fairly slim profit. Not many contractors, especially contractors on residential projects, include so small a profit in their bid. On extremely large projects such as highways, power plants or dams, the contractor may allow only 1% or less for profit ... especially if the amount to be received is based on the contractor's actual cost rather than a fixed bid. Residential construction, especially reroofing work, traditionally carries a higher profit margin because the size of jobs is much smaller than other types of work and the risk of significant cost over runs is larger. Probably 8 to 10 percent profit is a reasonable expectation on most jobs with very small jobs running to as much as 25%.

Of course, there is more to "profit" than just how much profit you would like to earn. Sharp competition will reduce the amount of profit you can figure into your estimate. If you include too much profit in your bids you will find yourself under-bid for the jobs you would like to have. If you have developed a specialty and you can do a particular type of work better than other roofers in the area and have enough work to keep you busy, then you may increase your profit by one or two percent or even more. Roofers who use modern sales methods have been very successful at maintaining high profit margins. When work is less plentiful many roofers take work at little or no profit margin in order to keep their best crews busy (and themselves in business).

In practice there is no single profit figure which will fit all situations. For most residential work an 8 to 10 percent profit is a very nice expectation. A roofer who has all the work he needs and wants and is asked to bid on more work may figure a 15% profit is not excessive. At the end of each year the profit on your business should give you a reasonable return on the money you have invested in the business (after you have taken a reasonable wage for yourself). You should earn a profit equal to 8 to 12 percent of the "tangible net worth" of your business. The tangible net worth is the value of all the assets of your business less the liabilities (anything your business owes) and less any intangible items such as goodwill, patents, or copyrights.

The small roofing contractor who has only a vehicle, some tools and a few hundred dollars working capital, may have a tangible net worth of less than $5,000. Any profit he shows, after he has taken a reasonable wage for himself, will very likely be used to buy additional equipment and increase his working capital. Still, he should include a profit in each job which will give him a return at the end of the year of between 8 and 12 percent of his $5,000 tangible net worth.

Crew And Equipment Management

Working a crew of roofers by the hour requires a lot of organizing and planning. Complete and thorough estimates will ensure that you deliver the right materials to the job site. Don't let your crews stop for lack of materials. Don't let your crews have the free use of major equipment without putting someone in charge of it. Someone should have responsibility for each piece of equipment. Make sure that the responsible individual knows the proper use and the safety requirements for each piece of equipment. Designate one person to be responsible for and to operate each truck. You or your superintendent should make regular inspections of the equipment and vehicles so as to be up to date on their condition. Keep the most necessary spare parts on hand in case of an emergency; V-belts, bearings, spark plugs and extra oil may be needed in a hurry. Maintain a weekly maintenance service schedule for lubricating bearings, cleaning spark plugs, cleaning gas filters and draining gas tanks and lines to remove any moisture. Set up a monthly service schedule to include such services as changing the oil, tightening the V-belts and examining all exterior moving parts. Maintain service records on each major piece of equipment and all vehicles. Don't allow yourself to get too busy to continue the records and maintenance services. Equipment breakdown can be costly and may cause a serious accident.

When your roofers work for a set price per square, the cost of labor can be estimated with a high degree of certainty. The profit you figure into each job should be the profit you end up with. But when your crews are paid by the square, you do not have as much control over their work. Until you are sure of the quality of their work you should inspect each job before they are paid. After a few jobs you can easily determine if you can afford to keep them on or if you should lay them off. Sometimes, if you are not careful, some applicators will use young, inexperienced roofers or will subcontract the work to others without your knowledge. This can seriously affect your reputation. Set down company rules and make sure that they are followed.

With most shingle applicators that work by the square, it is best to pay wages according to the total area the job is sold for. This will keep the roofers from trying to use up all of the material on each job. This way you should have more material left over. This type of working arrangement is used very successfully by many experienced companies.

If your applicators work by the hour, you should furnish the equipment necessary to cut costs and increase profits. Use roof loading equipment to get the shingles on the roof. Use laborers to distribute the material over the roof area and make sure the piles are not in the applicator's way. Some roofers use pneumatic staple guns in an effort to cut the cost of application. Balance this time saving against the additional staple cost. Power stapling might not be the answer for your particular situation. Some roofers and carpenters have a very hard time adapting themselves to the working position that is required when stapling. If you make the decision to purchase staple guns, follow up with a training program to get your applicators accustomed to them. Staple guns can save you money on new roofs but are not generally recommended for reroofing.

Training Programs

Regardless of how you pay your applicators, maintain performance records on each individual. This will tell you what each applicator needs to learn; sealing vents, making tie-ins or whatever. The foreman or whoever gives the training sessions can tell from the records exactly what should be covered. Safety suggestions should also be included in the training sessions.

A good employee-employer understanding as to what is expected will benefit both you and your crews. You expect fair treatment and you should try to give it. A company should retain only the type of applicators that will help build the company. Any other type of craftsman can only hurt you and your company. Even a small company can easily afford the best advertisement that there is: the recommendation of satisfied customers. Quite often a satisfied customer will sell your roofing jobs to a neighbor, a friend, someone at his work or a relative. Establishing a yearly roof inspection for your customers will bring you into contact with these leads as well as help build a good reputation.

Most roofing companies are either maintaining or building a reputation and are developing a public image to help increase sales. You have the right to request that your employees maintain good personal and job appearance because your applicators and their work are on display constantly. As an employer your future and your income depend on your employees. Impress on your employees the importance of their appearance, their attitude and their work. Ask them to see themselves as your client sees them.

Assume you have a medium priced home that needs a new roof. You signed a $1,200 to $2,000 contract with a roofing company that you know very little about. On the day that the company said they were to start reroofing you see a truck arrive with the material for your roof. Without knocking on the door and checking to see if you need to get your car out of the garage, the truck backs into your driveway and rolls onto the grass, leaving large ruts. A "professional craftsman" gets out of the truck, looking as though he is just getting off work instead of going to work. He casually unties the heavy wooden ladder, walks through your flowers and slams the ladder against the aluminum gutter. Next he starts loading the shingles onto your roof. Instead of bending over slightly and easing the bundles down onto the roof, he drops the bundles, causing the whole house to shake. As a result of thoughtless acts like these, many roofers are run off before they can get started. These are not imaginary situations or happenings but have occured far too frequently for many companies. If the roofers are not stopped, your client will get his roof but will never forget the rude treatment he also received. Satisfied customers are the cheapest and the best advertisement that a company can get. Neighbors are very good potential customers. A good roof, a satisfied customer and a professional crew are excellent salesmen. They are something that every roofing company needs to stay in business.

As your crews are installing shingles, have them maintain a neat jobsite appearance. Load the trash into the truck or pile it up neatly as it accumulates. Don't let the shingle wrappers blow all over the neighborhood. Don't allow scrap and shingles to slide or drop down onto lawn furniture, awnings, flowers or the front porch. Little things like this will protect your job and create satisfied customers. Your applicators' attitude also makes a big difference. On many jobs your roofers will arrive at the job site to discover that the roof is steeper than they anticipated or that it is very cut-up. Consequently, they waste a lot of time before they begin work. If they are well-motivated and trained, they will go to work immediately and overcome the unanticipated problem. A poorly trained and motivated crew would walk around the house several times, drink some more coffee and then go to work with a half-hearted effort. This type of attitude will end up costing you a lot of money. Roofers cannot expect an easy house every time. Every so often they will have to take on a tough job. Train your crews to understand that the best thing to do is simply go to work and get the job done so that they can then get on to an easier job.

The good roofer who is neat and clean and makes your customers happy should be used on your more important jobs. Many companies pay roofers like this a little more so that they will stay with the company.

## Starting A New Company

If you're setting up a new roofing company, there are a number of steps you have to take before you are in business. Starting a successful roofing business takes planning, skill and no small amount of perseverance. To make your planning easier, the following checklist will take you through the essential steps to opening your business. The phone directory will put you in touch with the city, county, state and Federal offices that must be contacted.

## 8 Weeks Before Starting

Be sure you have any contractor's license you need to operate in the states where you plan to do business. It may take 8 weeks or more to get your license if any examination is required.

## 4 Weeks Before Beginning

Open the business bank account and have checks printed with the business name. If you plan to operate as a corporation or a partnership, prepare the necessary forms or have an attorney prepare the documents for you. If you are going to operate under a fictitious name (a name other than your own name or the name of your corporation) file a fictitious name statement with the county where your business is located. Publication of the name may be required and this can take up to a month. Obtain an Employer Identification Number from your local Internal Revenue Service office.

2 Weeks Before Beginning

Apply to the appropriate state agency for unemployment insurance coverage. This must be done before you hire your first employee. Contact an insurance agent for the insurance coverage you need (worker's compensation, liability, fire, etc.). Request a business phone and other utilities that will be necessary. You will probably have to pay several advance deposits.

1 Week Before Beginning

Apply for the city or county business licenses you will need. You may need a resale license also. Check with the state authority administering the sales tax laws. A bond or a cash deposit may be required. Arrange to have business cards, stationery, invoices and advertising circulars printed.

Once you are in business you have passed the first hurdle. Building sales will be your major concern for a while. Chapter 13 is designed to help you sell the services your roofing company offers.

# Chapter 13
# Selling Your Services

It is not enough to be a good roofer if you want to make a good living in the roof business. You must also be a good salesman of your wares, and of yourself. In time, of course, your reputation as a roofer can be a valuable sales asset. But when you're just starting out you don't have a reputation. Moreover, even after your business is well established, a good reputation alone won't be enough to keep your firm growing in a field as competitive as roofing. You will still have to keep on selling and advertising your firm and yourself, or get someone else to do it for you. As a matter of fact, if you are unable or unwilling to do the sales work yourself, you probably will have to hire someone to do the selling for you.

If you specialize in residential work, you may find that an aggressive advertising campaign and continuous sales activity are needed to maintain a good sales volume during all seasons of the year. These selling efforts, of course, must be matched by a reputation for good work and honest dealing. Your firm doubtless will get work through contracts won in competitive bidding. General contractors will have you on their lists of bidders, and some customers will come to you, particularly after you have built a reputation. However, even if you get most of your work through competitive bidding, you will need to do some active selling to keep your crews busy during slow periods. Today, about 75% of all shingle roofing work is reroofing for homeowners and the amount of reroofing needed is sure to increase in the future. If you plan to capitalize on this growing market, you should know how to advertise and sell your services.

Your Advertising Budget

It pays to advertise. You've heard that often enough. But the question is, how *much does* it pay? Or, looking at the other side of the coin, how much should you spend on advertising?

A figure of 2 percent of gross sales is frequently quoted as a guide. For a roofing company doing $200,000 in business a year, this would provide $4,000 a year for advertising expenses. In practice, though, few roofing contractors grossing $200,000 a year spend that much. There is a strong temptation to scrimp on the advertising and promotion budget.

This is particularly so for a new company which usually has limited capital. But if yours is a new firm, you will need a well-planned campaign to sell your services, even more than an established firm does.

Advertising is simply a form of selling by means of publicity. It makes the reader familiar with the name and the service featured, associates that service with the name, makes personal sales efforts more productive and creates good will. It proves an economical means of keeping in touch with clients when it would be impossible to do so by personal calls. It secures inquiries and frequently results in sales direct from the advertising.

Too many of us are inclined to look upon advertising as a means of securing business when we have run short of contracts to be carried through or experience a temporary lull. It is easy to emphasize advertising as one part of selling and overlook its broader advantages -- reaching out to thousands of people with the personal message of a salesman and roofer, keeping clients sold and attracting others. Advertising is often used in a hit-or-miss fashion and the sales results are equally irregular. Regardless of how much or how little advertising you may do as a roofing contractor, you will find that to be successful, the various forms of advertising must be used with some judgment and follow a well understood purpose.

Planning Your Advertising

You must arrive at a decision about the type of client you want. What class of work are you anxious to secure? Will you build a reputation for high class work, or high-speed at low cost? Or will you go after any class of work that may come your way? Will you accept repair work, or will you limit yourself to new contract work? In what way does your service differ from that of your competitors?

Considering the class of work you prefer, what are the possibilities in your vicinity? Are your opportunities purely local, or do they extend over a large surrounding area? How will the seasons affect your work locally?

Has the market for roofing been limited in your territory for any reason in the past? Are there people who can use your services who are not aware of this fact? Are there sections where opportunities can be created for your class of work? What is your competition? How strongly established is it? Does competition offer what you have to offer -- the experience, the training, and familiarity with materials and newer construction ideas? How does the quality of their service compare? These and other questions along similar lines will be of material help to you.

Familiarize yourself with any advertising that has been done by other local roofers. Keep a file of their advertising as well as your own. Learn, if possible, what results they secured. If unsatisfactory, what was the reason? This alone will open an avenue of inquiry that should prove profitable to you as a guide in your own plans. Ads that appear again and again are successful and should tell you something about what works best.

Learn through the business papers, through manufacturers' salesmen, and other channels, what publicity or advertising plans and stunts have proven successful for other roofing companies, no matter where they may be located. Possibly the ideas can be adapted to your own advertising. Take the lumber yards, the building material people -- their problems are similar to your own, and their experience will give you valuable suggestions. The principles behind the advertising of allied lines are closely related.

Just as in your sales work, you should try to imagine yourself in the place of your prospect -- think in his language and from his side of the fence. He wants to know in what way you are going to benefit him, and why he should favor you above someone else. When reading your advertising he will respond to suggestions of self-benefits -- the same principles as apply in selling.

Effective Forms Of Advertising

Naturally you would hardly expect to use every form of advertising that is available. Select the form of advertising that best suits your community, advertising budget and type of business. Try to visualize who your prospect is and what type of advertising will reach him most economically:

Newspapers - advertisements
Newspapers - press notices and interviews
Billboards - posters
Calenders
Bus stop benches
Letters
Advertising matter
  Cards, folders
  Match books
  Pencils, rulers
  Booklets
  Broadsides
Advertising novelties
Photographs
Sponsoring a bowling team

Newspaper Advertising

Newspaper advertising is the mainstay of those whose work is concentrated in a local territory. The value of the newspaper lies in the fact that (a) it carries news and ensures concentrated circulation, (b) it reaches the people often, (c) it reaches into the home.

Naturally, its effectiveness depends on what you put into the space you use.

You will find ample material for copy or advertising material in the work you are doing. Be specific in your statements. To announce that your service is "the best that is offered" is not strong advertising. But the fact that you reroofed "five ranch style homes in record time" would be news of interest to other property owners. What have you been doing for others that would prove interesting? What feature of your service would influence business? What local event or situation presents an opportunity for a tie-in to your services? If you think of nothing else, go back to your analysis of your services.

Writing the copy for your advertising is largely a matter of using your good common sense, a matter of talking to an individual through the printed word instead of by word of mouth. Forget the fine language. Talk the good, homely Engish you would use in writing a friendly letter. In fact, advertising experts agree that when writing copy you should try to picture in your mind some one individual, such as a property owner, and write the copy as if you were speaking to that person.

Try to forget that what you are writing is to go into print. Just call to mind what you would like to have those readers know. Just imagine each has a job you would like to handle, and tell them why you should handle it. You'll be surprised at the sincere, straight-from-the-shoulder appeal that results. Brush it up a little later if you wish, but not too much. Remember, it is the homely, expressive, everyday phrase that gets under the skin and gets action.

Change your ads frequently. Indifference in this respect creates a bad impression and is rather expensive. This is especially evident where copy is run that no longer applies to current conditions. For example, copy emphasizing the importance of inspecting a roof before the rainy season begins should not be run so late in the season that those planning such work have already arranged for it.

Keep up to the minute on materials, prices, labor conditions. Frequently you will be able to feature reductions in costs which will bring many people into the market who have held back in the hope of just such a change. This gives you the reputation for being "on the job", well posted, thorough, and a good man to call in when estimates are wanted.

Use enough space in your newspaper to tell your story without unnecessary crowding. Use it frequently enough so you will not be forgotten.

Otherwise people may get the impression you are out of business. Use copy that appeals to the prospect. Make your advertisements attractive by the use of illustrations, if possible, and by using plenty of white space around the type to set it off. Large space is not necessary. Many a well planned campaign has relied on ads ranging in size from a few lines in the classified single newspaper to ten inches of double column in several papers. Much depends on the way the ad and the copy are laid out.

Figure 13-1 shows two possible display ad layouts. You should have no trouble adapting these to your needs or thinking up ads that will develop prospects for you.

Newspaper Publicity

Be friendly with the editor of your local newspaper. Make it a point to give him news items from time to time, but use a little discretion and tact -- don't make the reading notice a personal history.

As a roofing contractor and businessman you are a local institution serving the needs of your community. The public is interested in what goes on there. What are some of the things you are doing that would interest them? A new type of roofing material used on some merchant's store, a new roof on a factory building, an old landmark to be reroofed. If you are active in your local trade association, let the newspaper know when you attend a convention or meeting. It adds to your reputation -- and is news.

On special occasions there will be opportunity to give your opinion on matters of a local nature, or on general building conditions in relation to the city. An interview has great advertising value, keeping your name prominently before the public. All such advertising is valuable. Some say its more valuable than paid advertising.

Billboards, Posters And Signs

Moderate sized billboards or signs can be used to good advantage along major highways or elsewhere in districts especially active in building. Before placing such signs along the roads, however, it is best to consult the law of your state as to any restrictions or conditions. In proportion to the number of people who see these signs, this form of advertising is looked upon by many as one of the most economical forms you can use.

Do not overlook the opportunity presented right on the job itself. A sign large enough and attractive enough to be conspicuous should indicate that "So-and-So" is doing the work.

Such signs should be made up by a professional sign painter. White letters on a black background are considered the more attractive.

Whenever you do a roofing job in a residential area, it is a good idea to identify yourself to the neighboring property owners. There is no better endorsement of your service in a neighborhood than the fact that you put a good-looking, quality roof on the Smith or Jones house. The neighbors of Smith or Jones have roofs that are probably in about the same condition as the Smith or Jones roof. About the best and easiest way to identify yourself is with a door knob hanger as in Figure 13-2. Place about a dozen of these on front door knobs of the adjoining houses. You probably won't get many responses right away. But many of these hangers will be filed away for future reference. Before a year or two have gone by you should get several responses. A homeowner with a leaky roof needs professional help and he knows it. You may be the only roofer he knows anything about and no one wants to call a stranger when they have an expensive problem.

Match Books, Pencils, Rulers, Etc.

Despite the ever increasing use of "giveaway" items for advertising purposes, the fact remains that many of these are good advertising. They serve the purpose of keeping your name before prospective homeowners and other users of your services, and permit you to emphasize some feature of your service.

This form of advertising can be distributed through the mail with your regular correspondence or follow-up letters. For example, you could arrange with your bank to supply them with small calenders to enclose with monthly statements or to use on counters. A calender is effective because it is available for reference when hung up on the wall or placed on a desk. Naturally, a calender that is cheap and gaudy in appearance defeats its purpose for it generally finds its way into the waste basket. If you use calenders at all, spend enough to get good ones. Remember, a calender can be practical and attractive at the same time. Calender manufacturers can furnish remarkably attractive designs at a moderate price.

You will find that bus stop bench signs are good reminder advertising. It is a purely local medium, keeping your name before the reader. Seasonal appeals receive effective and quick response.

Letters

It is well to caution you here that if you believe that the advertising you do is going to sell roofing without any other support on your part, you are going to be disappointed. What it will do for you is secure interested inquiries from people planning to reroof their homes or having other need for your services. From then on it is up to you, personally. But this is all the average salesman needs. The letter, as part of your advertising plans, should be looked upon as the very foundation of your activities. True, in sending it to a large mailing list it is reaching a great number who are not at the time in the market for your services. But you will find that out of every thousand people you reach, there is a certain proportion who are in the market. Your letter, if effectively written, will secure replies from a paying percentage of these.

Naturally, a personally dictated or written letter might be sent to those prime prospects that have houses that will need new roofing in the next few months. But when you address a large number of prospects, such a letter is impractical. You must use a printed letter which is nothing more than many duplicates of the same letter. The names of those to whom it is sent can be filled in at the top on a typewriter. With such a letter going to a home, there is little or no difficulty getting it to the right people. But when it is sent to a business office -- passing through several hands -- the problem is to get it to the right individual. Then comes the problem of getting it read. If you know the name of the individual, you can address him personally. Getting the letter read is influenced by the appearance of the letterhead, the type arrangement, the length, and the specific appeal of the letter itself. Simplicity and directness should be the keynote throughout.

The chances for a reply are influenced by the directness of the message, short interesting paragraphs, a sensible, readable approach or message, and an ending that logically calls for a reply.

There are certain practices in letter writing that have proven effective and are worthy of consideration. Not that such letters will infallibly get business, but they have more of a fighting chance. These rules are summarized as follows:

1. Get to your subject at once; don't fence; talk to your prospect.

2. Never make an assertion that is debatable or untrue.

3. If you make a claim, prove it.

4. Never interrupt the steady flow of your prospect's thoughts by dragging in irrelevant ideas.

### What's 10 years old and has middle age spread?

Could be your roof.

Ten years is a long time for your roof. Past middle age when you realize a roof usually lasts about 15 years.

A roof check can tell if there are any loose or missing shingles, dried out caulking or rusted flashing. It could mean extra years of service for your roof. Finding a potential leak before it starts could save you hundreds of dollars in repairs to your home.

Call us. We'll be glad to check your roof and make any needed repairs reasonably. When it's time for a new roof let us tell you why the best roof is your best buy.

Contractor name

Address

City     State     Zip

Telephone

**Sorry!**
Please excuse the noise as we top off

the _____'s

house at _____
with new roofing shingles.

Stop over and see your neighbor's new roof, it's a beauty!

- - - - - - - - - - - - - - - - - - - - - - - - - - -

We would be happy to show you how to spot potential trouble on your roof and what to do, fill out the card and mail today.

Roofing Company
Address
City State Zip

Name _____

Address _____

City _____ State _____ Zip _____

### A roof, is a roof, is a roof!

Poets we're not. Roofers we are.

What we'd like to do, now during our off season, is inspect your roof. We'll check for loose or missing shingles, dried out caulking and rusted flashing.

If your roof is more than a few years old, chances are it could use a going over. If repairs are needed, we'll give you an estimate on doing them, quickly and reasonably.

Contractor name

Address

City     State     Zip

Telephone

## Selling Your Services

5. Make your letters easy to read. Use short sentences and paragraphs.

6. In your letter, give or imply all the facts about your proposition that the reader wants to know.

7. Avoid confusing the prospect by presenting several propositions from which he must choose.

8. Make your letter portray advantages to be gained, instead of evils to be avoided.

9. Give a logical reason for a reply.

These suggestions apply to the preparation of a personal letter as well as a form letter. Be human, be natural, be brief. Eliminate the "hot air". Nobody believes it anyhow, and it simply fills valuable space that could be used for a real sales point. As a reason for a reply, suggest a benefit to be derived. A postage paid business reply mail card should be included to make a reply easy and painless.

### Printed Matter

Small printed folders, circulars and envelope stuffers (small circulars) to accompany letters can be effective. They can be used to feature some particular job you recently completed, or to emphasize a class of work that you are especially anxious to secure.

Suppose you plan a small four page folder the size of an envelope. The first page could carry a descriptive title, the two inside pages an illustration of some of your completed jobs with descriptive copy, and the last page a summary of the services you offer with possibly a testimonial from a client.

The advantage of a circular with a letter is that it permits you to tell your story in the circular, leaving your letter to be devoted to selling your services.

Advertising will not accomplish the impossible; nor can it make a permanent success of any service that in itself is not meritorious. Nor can you expect poor advertising to accomplish what well planned advertising will do. Remember that in starting an advertising plan you set a standard of practice. You are committing yourself to maintain a certain level of performance. You are creating a reputation to which you must conform. This responsibility, this standard you set up, brings with it the reward of character, reputation, constantly increasing business, independence and other things we associate with success. In advertising, as in other business activities, reputation is the foundation of success.

### Follow Up Your List Of Prospects

Keep a record of prospects developed through the methods outlined here. Build up as large, and as complete a mailing list of prospects as possible. Carefully keep the name of every man and woman you have any hope of ever doing business with at any time. Keep a list of your past customers. Some roofers report that old customers are the best source of prospects. Send old customers a note at least once a year.

Business will not come to you without invitation. If you want more business, you must go after it, work for it, earn it. Use your imagination in developing possible outlets for your services. Form the habit of picturing every home in your community as a potential reroofing job.

True, all this means a little added effort on your part, and you may feel that you have too many other things to attend to. If that's the way you feel, just think back over the various successes you have heard of in the building industry. Who are the successful builders? Who are the men pointed to with pride as accomplishing great things in the industry? Need we answer these questions?

Summed up, it means simply this: If you want to get away from price competition -- a situation found in every industry -- you must give enough time to developing business from a creative angle. You can then make your estimates of cost exactly what they should be rather than cutting bids to meet the competition of some vicious price cutter. Remember that when a job comes to you unsolicited and without any logical reason for you being favored, you can rest assured that you are going to find it necessary to cut the corners in your estimates.

On the other hand, if you have created the work, if your ideas have been responsible for developing a job where there was none before, you are entitled to receive your full price, both for the time you spent in creating the business and in selling the idea to your client. As a general rule, there is little difficulty in getting your price for such contracts.

### Selling The Prospect

In the business of roofing, modern sales methods are just as necessary and effective as in any other industry. It is not enough to understand shakes, shingles and tile in order to go into the business. Roofing contractors must know how to sell. In fact, whether you are going into business for yourself or intend to merely sell your services, you should know something of salesmanship.

At the beginning you should know that the old idea that selling is a mysterious gift no longer holds true. While some men appear to be "natural born salesmen", there are certain principles which may be learned and followed by anyone willing to apply himself.

The chief problem in selling is to bring the one you are trying to sell (the prospect) to feel the way you do about your proposition -- to make others think as you do about certain things. The most practical way to do this is to use good, common sense, plain honest talk, and give every evidence of a sincere desire to solve the problems and serve the prospect. It is a matter of rendering personal service, and not necessarily a question of a highly persuasive personality.

### Routine And Creative Salesmanship

Selling itself may be divided into routine salesmanship and creative salesmanship. The former is simply a matter of filling an existing want, as when a property owner telephones you and asks the cost of putting on a new 3-tab roof -- and your price, without further comment, gets the job.

The creative salesman would drop in on a property owner, show him a modern, attractive colored and textured strip shingle adapted to his property, point out how it would increase the property value and add materially to the appearance. Closing such a sale would be creative salesmanship. In short, it is the difference between handing out what a man asks for, and in educating him to want the thing or service you think best for him.

Thus the true salesman creates wants -- wants that are so insistent and impelling as to overcome any other want at the moment. It is done by talking in terms of the prospect, putting your proposition in his language.

People do not buy shingles as such, but rather they buy benefits. A man does not buy an automobile. He buys travel, convenience, the open road, ease of operation, low upkeep, satisfaction of vanity. He does not buy a vacant lot so much as he buys a potential opportunity, enhancing land values, or a prospective homesite. Nor does he buy just shingles and your service to install them. He buys the security and piece of mind that a quality roof provides. He buys the admiration of his neighbors and friends who appreciate a property improvement as obvious as an attractive roof cover. Everyone buys what they expect to derive from the tangible thing secured -- the service. And so in selling a reroofing job, the sale is made in the mind of the prospect, not in the mind of the salesman. The salesman has simply supplied certain ideas which offset previous ideas or objections that were in the mind of the prospect. These newer ideas removed the objections that had been a bar to the action suggested.

### You Sell An Essential Service

Before you really can go to a prospect to secure a contract, you should convince yourself that you are really selling him something that you, yourself, would be proud of. Have a high regard for your occupation. Remember that a roofing contractor supplies one of the three essentials of life: food, shelter and clothing.

After we have satisfied ourselves that we really have something to sell or a service to give that is worthwhile, we then come to the point of satisfying ourselves on the question, "Why can I give this service better than the other man?" or "Is there any way that I can give this service better than my competitor?" If you realize that the other man can put on a better roof and do it cheaper and give better service to the prospect than you can, then it would be up to you to equip yourself so you could give a similar service or get into another line of work where you could give reasonable or better service.

In order to reach the point of having sold a prospect, there are five things that must be brought about in the prospect's mind. You must first secure his attention; then you must develop this into interest in your proposition; from here the prospect must be brought to believe in the value of the thing offered; then you must arouse a desire to possess; and finally you must bring this desire to a decision to buy and take the necessary action.

Before this development has taken place the prospect must be approached and the matter introduced to his attention. With this in mind we can actually divide the complete selling process into the pre-approach, the approach, the interview (or presentation), and the close. Throughout these four steps various arguments are mingled with proofs, objections, human appeal and countless other points of discussion.

The pre-approach is simply the preparation for the interview. It means getting as much information about the prospect and his side of the interview as possible, in advance of the call. Often a salesman or builder will go to see a prospect "cold turkey" -- that is, without any data concerning the prospect. It is best to secure all the information possible in advance. In the case of selling the idea of reroofing a house to a property owner, know something of his circumstances, the type of roofs on the better homes in the neighborhood, whether his wife will be deciding, where he can get financing, etc. It goes without saying that you will be thoroughly posted on your proposition from your side, ready to modify your arguments, meet objections, and maneuver around a cold turn-down.

Approaching The Prospect

Even your approach is a sale in one sense, for you must "sell" your prospect on the idea of giving you a certain amount of his time. Many people tend to shun the approach of anyone trying to sell them. You will find it almost fatal to say frankly: "My name is Mr. Brown. I'm a roofer, and I want to interest you in applying some new shingles." In fact, what answer could a man give such an approach other than "not interested?"

The approach that gets you somewhere in the shortest possible time is one that gives the prospect some idea of what you have to tell him in detail, but in terms of what it will accomplish for him while arousing his desire to know more. A little careful thinking and planning will help you to do this. A knowledge of the prospect and his business and a knowledge of what your proposition will do for him is needed.

Take the salesman who wants to sell you some shingles his company makes. Does he come at you with, "I'd like to sell you the shingles you're going to need on the job"? Not at all. He eases around with the suggestion: "Suppose I could show you a way to save about 25% on the cost of that roofing job, with a better piece of work and a more lasting job -- would you be interested?" Naturally! That is his approach, and he certainly sells you on the idea, for you want to hear more.

To omit this first step -- the approach -- is fatal to the rest of the interview, for on it is built the first basis of confidence and understanding -- the willingness to listen. The approach is based on self-interest. What will this do for me? What will it do for those close to me? So in order to approach your prospect to advantage, ask yourself this question: Why should this man give me the desired fifteen minutes of his time? Your honest answer will give you the key to your opening approach.

Apply this question to the reroofing project previously referred to. What would be your answer? Wouldn't it be, "Because I know such a roof would add to the appearance of this man's property and the type of shingle I have in mind would increase the resale value of the property far more than the cost. There's your opening approach in your answer. You are going to sell him because what you suggest will solve his problem, a leaky or worn out roof, and will benefit him, and not because you are a fellow lodge member, or wear the same color tie. You are going to make his property more useful, attractive and valuable, and get him interested by telling him in a way he can believe and that will quickly arouse his interest.

Representing Or Demonstrating Your Service

Having sold your prospect on the idea of listening to your story, you are ready for the presentation or demonstration. Bear in mind during your presentation that there are two avenues of appeal open to your use. One is an appeal to the prospect's reason and intellect through logic and sound argument; the other is an appeal to the prospect's emotions and imagination by positive suggestions and samples, illustrations or word pictures.

Few of us decide a matter by logic alone. In fact, there is a mixture of both reasoning and emotion, although it is surprising the number of sales made on the emotional appeal alone. By this we mean the introduction of a suggestion into the prospect's mind around which he constructs a mental picture in which he is the central figure.

Another point to bear in mind is that a human being cannot be influenced except through one of the senses of sight, hearing, taste, smell, and touch. And every idea or feeling that enters the mind creates a reaction. These reactions are expressed in the face, hands, eyes, shoulders, or other parts of the body in some way. These expressions tell the experienced salesman what is going on in the prospect's mind and indicate whether the right line of attack is being followed. So note carefully all reactions to your arguments or suggestions as you go over them with the prospect.

People understand pictures more quickly than words or verbal description. They are direct, simple, uninvolved and concrete, hence, use words and phrases which bring up pictures of the idea to be conveyed. They work on the imagination more quickly and effectively than high sounding terms, and it is this very word picture that we wish to establish in the prospect's mind. And a sample shingle or color picture is even better than a description. It claims the attention and holds it, allowing the mind to grasp something definite and to picture it in its relation to the prospect. It keeps the mind from wandering and focuses on one point. When we have laid before him a sample or full color brochure, his mind immediately registers "yes" or "no" to the idea of color, texture, etc. These points immediately become matters of decision, aiding in another step toward something definite.

Your prospect probably never thought about his roof very much. He knows its supposed to keep the water out and probably, for some reason he isn't quite sure of, didn't keep the house dry in the last rain. He doesn't really want to have much to do with the roof. Its a source of annoyance rather than pleasure and pride. As a salesman your job is half done if you can convince him that his roof is an important decorative element in the design of the house and that roof color, style, and texture as well as durability are important.

Let your prospect explain the trouble he has been having. Then, explain that most shingle and built-up roofs are designed to last about 15 years. Ask him when his roof was installed and whether he has had any problems before he called you. Your obvious concern for his problem and his willingness to discuss it establish a professional — client relationship that makes, when you finally give it, your opinion more valuable. Go with your prospect to inspect any leaks from the interior of the house. This is your prospect's real problem and he will want to show you what happened and emphasize how serious it is.

Next, go up on the roof to make an inspection. Take your prospect with you if he is willing to follow you and the pitch is not too steep. Take a pencil, tape measure and note pad with you. When you reach the roof, emphasize two things at once. Walking on a roof is both dangerous, because of the possibility of a fall, and hazardous to the roof. Foot traffic on a roof will only aggravate any problem the roof already has. Your prospect will get the point: A roof is no place for an unskilled amateur. Now begin your written evaluation of the roof. Make notes about each of the 7 problem areas listed below. Be thorough. List the good points as well as the bad. It establishes your credibility. Be honest. You're sure to loose the job if your prospect suspects your truthfulness. Be professional. Your prospect wants to believe that you are a real "pro" in the roofing business and that you know what you are talking about. Don't disappoint him.

*General roof condition.* Look for missing, blistered, split or curled shingles. Explain that shingles can be replaced inexpensively in a new roof but in an older roof it is easier to replace the entire roof rather than make piecemeal repairs.

*Flashing.* Look for rust spots and loose areas. Explain that rusted areas should be cleaned and painted. Metal that is rusted through should be replaced. Check for dried out caulking and emphasize that dry, stiff caulking should be removed and replaced.

*Loose and missing nails.* Look for nail heads that have torn the shingle or are no longer tight against the shingle. If you see poorly nailed shingles, comment on it but don't dwell on it. Your prospect will note that you recognize substandard practice but you won't gain stature in his eyes by criticizing whoever installed the existing roof, especially if he installed the roof himself.

*Worn spots.* Check the roof for patches of dark gray or black where granules have worn off. Also, check gutters and the area around downspouts for concentrations of loose granules. Explain that large amounts of loose granules are the best indication that the roof has lost its ability to withstand further sun, wind and moisture.

*Valleys.* Be especially aware of these critical areas. Look for torn and worn spots. Explain that leaks caused by a faulty valley can appear anywhere and should be corrected at once.

*Gutters.* Explain the importance of good gutters and that water from one slope should never be allowed to drain onto a different lower slope. Explain that leaves and debris in a gutter can cause water to back up and soak under shingles.

*Roof attachments.* Look for signs that anything has been attached to or nailed to the roof. Explain that a T.V. antenna anchored to a vent without proper guy wires will vibrate and loosen caulking around the flashing.

Now give a well considered evaluation of what is needed. Don't recommend a new roof if one isn't needed. Go ahead and recommend minor repairs if that is what is called for. Explain that temporary repairs will be just that, temporary. Moreover, you can't guarantee that a similar problem won't develop next season. If you recommend repair rather than replacement and you do a good job of making the repair, you can be sure of getting a call when replacement is necessary.

If replacement is necessary, measure the roof while you are there and make a sketch to use in preparing the estimate. Explain why replacement is necessary and use your notes to make sure you cover every point. Make a sketch, if you want, to show how water can enter a roof at one point and run under shingles, felt and sheathing for some distance before leaking into the house. Explain that it may be possible to add a new roof without removing the old roof and that this will both save money and provide better roof insulation when completed. Don't make any promises about the sheathing if the roof is badly deteriorated. If a tear-off is required, make it clear that you can not evaluate the cost of replacing sheathing until it is exposed. If the flashing is in good condition, explain that it can be cleaned, painted and reused.

Once your client accepts your opinion that a new roof is needed, your job of selling roofing really begins. Anyone can sell a no-frills, minimum quality roofing job on a leaky roof in the middle of the rainy season. As a roofer you know that there is a trend away from the 240 pound basic white or dark shingles and to heavier weight and fiberglass textured shingles in brown and earth colors. Point out an example of the newer shingles on one of the better homes in the neighborhood. Some of the asphalt shingles have heavily textured surfaces and look like wood shingles. Explain that many of the newer shingles not only look better, but wear better, some being guaranteed for 20 or 25 years. Modern asphalt shingles carry a U.L. class "C" or class "A" fire rating and are wind resistant. In short, in the last 15 or 20 years there have been many improvements in roofing shingles that make it sensible to modernize the roof. Reroofing with the newer shingles will improve both the appearance and resale value of the property. Recommend one particular high quality shingle and offer an opinion on what color goes well with his house.

Now is the time to bring out some samples and pictures. Nearly all manufacturers of roofing products offer full color pictures and samples of their products. Many general circulation magazines carry full page color advertisements placed by roofing manufacturers featuring large, neatly landscaped homes with attractive shingle roofs. Have some of these and some color samples available so your prospect can visualize the new roof on his home. Stand back and let your prospect compare shades and textures against the house. Remember, your prospect will buy the admiration of family, friends and neighbors. He will buy security and piece of mind. He is *not* going to buy just a set of shingles. He is going to buy because he gets good value for his dollar.

Your first samples and pictures should be of top - of - the - line shingles you recommend. Point out the rich texture and deep shadow lines. Have your prospect pick up a sample and feel the heavy weight. Emphasize that these better shingles carry a 20 or 25 year guarantee rather than the usual 15 year guarantee. Then offer to figure an estimate for a reroof job using these better shingles. As you are figuring the cost, make these important points about top - of - the - line shingles. A good 25 year roof costs only about ⅓ more than the least expensive asphalt shingle roof but is guaranteed to last nearly twice as long. More important, many buildings will support only one additional roof before the entire roof surface has to be removed. The first reroof should be done with a very good shingle. Finally, emphasize that the better shingles just look better and will continue to look better for many years.

You know that selling better quality shingles increases your volume and profits. Your client knows this too. As a salesman you want to convince him that the better shingle offers the benefit of better appearance and lower cost per year of useful life and that these benefits outweigh the higher initial cost.

If your prospect balks at accepting your suggestion, offer to refigure the job based on lower cost, standard quality shingles. Be honest in explaining the difference in durability and weight between the better quality and lower quality shingle. Explain that the labor cost is nearly the same. Only the material cost is different.

Answering Questions And Objections

Throughout the demonstration and selling talk, many questions will arise which must be disposed of to the satisfaction of the questioner. Go over the matter carefully again. Remember, it is not enough that you understand how foolish the question is. Your prospect must be made to understand what he asks about. For it is in his mind that the sale is being made, not in yours. If you do not offset every objection he makes there is no sale.

These "objections" are often merely honest doubts on points not clear to him. You must supply the facts and truths to offset these ideas in order to bring about the sale. The best way to overcome these objections is to anticipate them whenever possible.

The prospect may object that he cannot finance the type of roof you recommend. If you have learned everything possible at the outset regarding your prospect, you have anticipated this objection by pointing out during the interview that the only cash outlay would be approximately so many dollars, while your bank would take care of the remainder.

Have your selling points marshalled and under control; be prepared to answer any questions that arise. In the event of the price question coming up before you are ready for it, try and defer this until later. Some salesmen, answering this point will say: "Why not see first whether this is what you want, then with the actual requirements before us we can figure more accurately how our costs will run. And I'm sure I can bring it to a basis agreeable to you." Then they will fall back to the point in the demonstration where they left off, and proceed.

Objections offered may embrace price, or service or countless other factors. Frequently, they may not have anything to do with the merit of your service. For example, when a prospect says he was once disappointed in the completion time promised by some contractor in the past, this momentarily becomes the leading consideration. Clear this up in its suggested relationship to yourself and then revert back to the main considerations.

When the resistance of your prospect begins to ease up, when he has pretty well agreed with your ideas as embodied in the plan, and starts asking questions of a supplemental nature, it is time to stop your demonstration and get down to closing.

## Closing

It is not the wisest thing to formally ask the prospect to sign a contract at this point. Many secure this same result by taking the willingness for granted and bring it up by test questions. For example: "I am just finishing up a job on the west side of town and could get my men started on this work by the first of the month. Would that be soon enough?"

If he balks, try and find the reason, then center your conversation on this point until it is eliminated. You may wish to show why *now* is the best time for a decision, presenting sound logical reasons for your advice. Show him why, as a matter of self interest, he cannot afford to delay or arrive at any other decision -- that to do so would reflect on his good judgment and opinions previously given.

He may want to "think it over". Then point out the result of doing so -- often a decision to wait a while results in disappointment later because he failed to decide. Add that the price you quoted is based on current labor and material costs and that you can not guarantee the quote beyond a reasonable length of time. Find out if there are any lingering doubts, any information you have failed to supply. If the demonstration has been complete, there is no reason why he should not proceed, if finances permit.

Often a man will say "no" to a proposition when he really does not mean it. He is actually waiting for additional facts, further reasons to justify a definite order. Many salesmen accept "no", only to learn that later the man said "yes" to an identical proposition. Supply facts, test for the reason behind the refusal. It may simply be a "stall" and can be made a logical starting point for a newer, well-directed sales effort. This is far more satisfactory than being put off, or "stalled".

You may have a prospect who has listened carefully. He may have agreed that what you say and demonstrated are true -- but still no sale. "You certainly have an attractive proposition, but I can't go into it right now; drop in Tuesday, will you?" That sounds very nice, but -- it doesn't mean a thing more than the fact that the man is putting off a decision. It means you should go over your demonstration and find its weak spot, then find the reason for the delay, and center on that. "Thinking it over", "Talking it over", "Drop in later on"; too many objections at the end of the demonstration and close show a weakness in the presentation -- an absence of sufficient facts, forcefullness, earnestness and faith in your proposition. Prevent these "objections" by anticipating them. Eliminate call-backs by developing your presentation for a decision, not an opinion.

Suppose your investigation of the prospect has shown you that your recommendation would prove a decided benefit to him and that he was able to carry it financially. Failure to "close" rests on your own presentation of the proposition. Check it over, picture your interview, call to mind all the objections brought up. Did you satisfy him on all those points? This self-examination is the secret of success in future demonstrations.

## When To Excuse Yourself

And when your prospect agrees to your proposition and gives you the "go-ahead", *get out!* Except for the discussion of matters relating to your proposition, and a few courteous comments, leave when you have put the sale across. Don't talk yourself out of it; don't stay to discuss points that might lead to doubts and excuses -- and putting off what has been decided. Your sale is concluded -- you have finishing touches to put into the job just discussed -- logical reasons to excuse yourself.

# PROPOSAL AND CONTRACT

Date _____ 19 _____

To _____

Dear Sir:

We propose to furnish all materials and perform all labor necessary to complete the following:

_____
_____
_____
_____
_____
_____
_____

Job Location: _____

All of the above work to be completed in a substantial and workmanlike manner according to the terms and conditions on the back of this form for the sum of _____

_____ Dollars ($ _____ )

Payments to be made as the work progresses as follows: _____

_____
_____
_____
_____

the entire amount of the contract to be paid within _____ days after completion. The price quoted is for immediate acceptance only. Any delay in acceptance will require a verification of prevailing labor and material costs.

By _____

Company Name _____

Address _____

State Licence No. _____

"YOU, THE BUYER, MAY CANCEL THIS TRANSACTION AT ANY TIME PRIOR TO MIDNIGHT OF THE THIRD BUSINESS DAY AFTER THE DATE OF THIS TRANSACTION. SEE THE ATTACHED NOTICE OF CANCELLATION FORM FOR AN EXPLANATION OF THIS RIGHT."

You are hereby authorized to furnish all materials and labor required to complete the work according to the terms and conditions on the back of this proposal, for which we agree to pay the amounts itemized above

Owner _____

Owner _____ Date _____

# Contract Terms

The Contractor agrees to commence work hereunder within ten (10) days after the last to occur of the following, (1) the building site has been properly prepared for construction by the Owner, and (2) the materials required are available to Contractor. Contractor agrees to prosecute work thereafter to completion, and to complete the work within a reasonable time, subject to such delays as are permissible under this contract.

Contractor shall pay all valid bills and charges for material and labor arising out of the construction of the structure and will hold Owner of the property free and harmless against all liens and claims of lien for labor and material filed against the property.

No payment under this contract shall be construed as an acceptance of any work done up to the time of such payment, except as to such items as are plainly evident to anyone not experienced in construction work.

In the event that any conflict exists between any estimate of costs of construction and the terms of this Contract, this Contract shall be controlling. The Contractor may substitute materials that are equal in quality to those specified if the Contractor deems it advisable to do so.

Owner agrees to pay Contractor its normal selling price for all additions, alterations or deviations. No additional work shall be done without the prior written authorization of Owner. Any such authorization shall be on a change-order form, approved by both parties, which shall become a part of this Contract. Where such additional work is added to this Contract, it is agreed that the total price under this Contract shall be increased by the price of the additional work and that all terms and conditions of this Contract shall apply equally to such additional work. Any change in specifications or construction necessary to conform to existing or future building codes, zoning laws, or regulations of inspecting Public Authorities shall be considered additional work to be paid for by Owner as additional work. If the quantity of materials required under this Contract are so altered as to create a hardship on the Contractor, the Owner shall be obligated to reimburse Contractor for additional expenses incurred. If the Owner or agent of the Owner furnishes material or performs labor on any portion of the work in progress, any loss to Contractor that results thereof shall be charged to the Owner. Any changes made under this Contact will not affect the validity of this document.

The Contractor shall not be responsible for any damage occasioned by the Owner or Owner's agent, Acts of God, earthquake, or other causes beyond the control of Contractor, unless otherwise herein provided or unless he is obligated by the terms hereof to provide insurance against such hazards. Contractor shall not be liable for damages or defects resulting from work done by subcontractors. In the event Owner authorizes access through adjacent properties for Contractor's use during construction, Owner is required to obtain permission from the owner(s) of the adjacent properties for such. Owner agrees to be responsible and to hold Contractor harmless and accept any risks resulting from access through adjacent properties.

The time during which the Contractor is delayed in his work by (a) the acts of Owner or his agents or employees or those claiming under agreement with or grant from Owner, or by (b) any Acts of God which Contractor could not have reasonably foreseen and provided against, or by (c) stormy or inclement weather which necessarily delays the work, or by (d) any strikes, boycotts or like obstructive actions by employees or labor organizations and which are beyond the control of Contractor and which he cannot reasonably overcome, or by (e) extra work requested by the Owner, or by (f) failure of Owner to promptly pay for any extra work as authorized, shall be added to the time for completion by a fair and reasonable allowance. Should work be stopped for more than 30 days by order of any agency of government, the Contractor may terminate this Contract and collect for all work completed plus a reasonable profit.

Contractor shall at his own expense carry all workers' compensation insurance and public liability insurance necessary for the full protection of Contractor and Owner during the progress of the work. Certificates of such insurance shall be filed with Owner if Owner so requires. Owner agrees to procure at his own expense, prior to the commencement of any work, fire insurance with Course of Construction, All Physical Loss and Vandalism and Malicious Mischief clauses attached in a sum equal to the total cost of the improvements. Such insurance shall be written to protect the Owner and Contractor, and Lien Holder, as their interests may appear. Should Owner fail to do so, Contractor may procure such insurance, as agent for Owner, but is not required to do so, and Owner agrees on demand to reimburse Contractor in cash for the cost thereof.

Where colors are to be matched, Contractor shall make every reasonable effort to do so using standard colors and materials, but does not guarantee a perfect match.

Any controversy or claim arising out of or relating to this contract, shall be settled by arbitration in accordance with the Rules of the American Arbitration Association, and judgment upon the award rendered by the Arbitrator(s) may be entered in any Court having jurisdiction. Should either party hereto bring suit in court to enforce the terms of this agreement, any judgment awarded shall include court costs and reasonable attorney's fees to the successful party plus interest at the legal rate.

Unless otherwise noted in this agreement, the price quoted does not include removing or replacing fascia, trim, sheathing, rafters, structural members, siding, masonry, vents, roofing, caulking, metal edging or flashing of any type. If, during the course of work, it should become apparent that any such portions of the structure should be repaired or replaced, Owner may authorize Contractor to do such additional work and charge Owner for the additional labor and materials required plus a reasonable profit.

The Owner is solely responsible for providing Contractor prior to the commencing of construction with such water, electricity and refuse removal service at the job site as may be required by Contractor to effect the construction covered by this contract. Owner shall provide a toilet during the course of construction when required by law.

The Contractor shall not be responsible for damage to existing walks, curbs, driveways, structures, cesspools, septic tanks, sewer lines, water or gas lines, arches, shrubs, lawn, trees, clotheslines, telephone and electric lines, etc., by the Contractor, sub-contractor, or supplier incurred in the performance of work or in the delivery of materials for the job. Owner hereby warrants and represents that he shall be solely responsible for the conditions of the building site over which the Contractor has no control and subsequently results in damage to the building or injury to persons or property.

Contractor agrees to complete the work in a substantial and workmanlike manner but is not responsible for failures or defects that result from work done by others prior, at the time of or subsequent to work done under this agreement, failure to keep gutters, downspouts and valleys reasonably clear of leaves or obstructions, failure of the Owner to authorize Contractor to undertake needed repairs or replacement of fascia, vents, defective or deteriorated roofing or roofing felt, trim, sheathing, rafters, structural members, siding, masonry, caulking, metal edging, or flashing of any type.

Contractor makes no warranty, express or implied (including warranty of fitness for purpose and merchantability). Any warranty or limited warranty shall be as provided by the manufacturer of the products and materials used in construction.

Owner hereby grants to Contractor the right to display signs and advertise at the building site.

Contractor shall have the right to stop work and keep the job idle if payments are not made to him when due. If any payments are not made to Contractor when due, Owner shall pay to Contractor an additional charge of 10% of the amount of such payment. If the work shall be stopped by the Owner for a period of sixty days, then the Contractor may, at Contractor's option, upon five days written notice, demand and receive payment for all work executed and materials ordered or supplied and any other loss sustained, including a profit of 10% of the contract price. In the event of work stoppage for any reason, Owner shall provide for protection of, and be responsible for any damage or loss of material on the premises.

Within ten days after execution of this Contract, Contractor shall have the right to cancel this Contract should he determine that there is any uncertainty that all payments due under this Contract will be made when due.

This agreement constitutes the entire contract and the parties are not bound by oral expression or representation by any party or agent of either party.

The price quoted for completion of the structure is subject to change to the extent of any difference in the cost of labor and materials as of this date and the actual cost to Contractor at the time materials are purchased and work is done.

# Notice To Customer Required By Federal Law

You have entered into a transaction on_____which may result in a lien, mortgage, or other security interest on your home. You have a legal right under federal law to cancel this transaction, if you desire to do so, without any penalty or obligation within three business days from the above date or any later date on which all material disclosures required under the Truth in Lending Act have been given to you. If you so cancel the transaction, any lien, mortgage, or other security interest on your home arising from this transaction is automatically void. You are also entitled to receive a refund of any down payment or other consideration if you cancel. If you decide to cancel this transaction, you may do so by notifying

_____
(Name of Creditor)

at_____
(Address of Creditor's Place of Business)

by mail or telegram sent not later than midnight of_____. You may also use any other form of written
(Date)
notice identifying the transaction if it is delivered to the above address not later than that time.

This notice may be used for the purpose by dating and signing below.

I hereby cancel this transaction.

_____           _____
(Date)                                     (Customer's Signature)

**Effect of rescission.** When a customer exercises his right to rescind under paragraph (a) of this section, he is not liable for any finance or other charge, and any security interest becomes void upon such a rescission. Within 10 days after receipt of a notice of rescission, the creditor shall return to the customer any money or property given as earnest money, downpayment, or otherwise, and shall take any action necessary or appropriate to reflect the termination of any security interest created under the transaction. If the creditor has delivered any property to the customer, the customer may retain possession of it. Upon the performance of the creditor's obligations under this section, the customer shall tender the property to the creditor, except that if return of the property in kind would be impracticable or inequitable, the customer shall tender its reasonable value. Tender shall be made at the location of the property or at the residence of the customer, at the option of the customer. If the creditor does not take possession of the property within 10 days after tender by the customer, ownership of the property vests in the customer without obligation on his part to pay for it.

# Roofing Estimate Form

Project _____  Sheet No. _____

Date _____  Checked By _____

| Description | Quantity | Unit | Material | Labor | Total |
|---|---|---|---|---|---|
| | | | | | |

# Reference Section

The information that is listed in this section has been condensed from manufacturers' brochures and is presented as an aid to product selection. The exact pattern of the shingles and the shake appearance may vary slightly from the drawings but the procedure for starting will be as described. To be assured that the warranty will be honored by the manufacturer, the roofer should follow the directions printed on the wrapper. Most manufacturers will supply you with more information if it is needed. Due to limited space in this manual, all of the products of each manufacturer are not described. Most of the newer composition shingles are included, however.

Some of the shingles featured should not be applied to a roof slope if the pitch is less than 4 inches per foot. Consult the manufacturer or check the shingle wrapper before application if in doubt. For slopes less than 4 inches per foot but not less than 2 inches per foot, the manufacturers often suggest the following procedure: Apply a metal drip edge to the eave directly over the wood. Then apply a double layer of 15 pound asphalt felt as an underlay before shingling. Start with a 19 inch strip at the eave and keep it flush with the metal drip edge. Over this lay a 36 inch strip of 15 pound asphalt felt. The following strips of felt should overlap the preceding felt 19 inches each time. In areas where the winter daily average temperature is 25° F or less, or wherever there is a possibility of ice or snow building up at the eaves causing a back-up (freezeback) of water, proceed as follows: Cement felts to the starter strip and to each other (with plastic cement or the proper cold application adhesive) from the eaves to a point 24 inches inside the inside wall line of the bulding. The bundle wrapper will have more specific instructions for low slope roofs.

On mansard or steep roofs the following is appropriate. For slopes exceeding 60° or 21 inches per foot, six nails must be used to fasten each shingle in place. In addition, each shingle must be stuck down with plastic cement immediately upon installation. More complete information can be obtained from the manufacturer or the Asphalt Roofing Manufacturers Association at 757 Third Avenue, New York, New York 10017.

Most manufacturers print handling and storing instructions on the shingle wrappers. As much care should be given to handling and storing of the shingles as to the application. The shingles should not be stacked over 11 bundles high. They should be stacked on pallets no more than 2 high or something smooth and be elevated from standing water. On some job sites the wind may make it difficult to keep the shingles covered. The shingles should be covered with a waterproof cover

Many people may be involved in the handling and storing of the shingles. All should do their part. To insure this the roofer should consult with the builder, if he furnished the shingles, or the supplier. A simple phone call from the builder or contractor will probably help on the next job. In any case, someone is or should be responsible for the condition of the material. Also, note that some shingles have a color code number stamped on the wrapper. Batch coded shingles should not be mixed with shingles of a different code number. Often the shingles are stacked out in the rain and the code number becomes too faded to read. The result is a poorly matched roof. The quality of any product can be weakened by improper handling and storing and this is especially true of shingles.

In the descriptions that follow the nail requirements have been increased slightly to reflect the actual experience of most roofers. Some nails are lost and others are used to nail accessories and roofing felt. Consequently, slightly more nails are needed than are actually used to place shingles.

Every effort has been made to describe and illustrate the various products as accurately as possible. The information in this section is correct as of the publication date but no doubt will change in time as manufacturers make changes in specifications and design. The best source of current product information is always the manufacturer. Some of the shingles illustrated in this section show a factory adhesive. The exact size, shape, spacing and precise adhesive location may vary somewhat from the illustrations in this section.

# Bird Roofing Materials

BIRD AND SON, INCORPORATED
Washington Street
East Walpole, Massachusetts 02032

District Offices, Building Material Division
(Asphalt Roofing Plants and Offices)
Stark Industrial Park, Charleston, SC 29405
Pleasant Street, Norwood, MA 02032
Amboy Avenue, Perth Amboy, NJ 08861
Aero Drive, Shreveport, LA 71107
(Pacific Div. Office), 2555 Flores St., San Mateo, CA 94403
6350 NW Front St., Portland, OR 97208
110 Waterfront Road, Martinez, CA 94554
1430 "E" Street, Wilmington, CA 90746

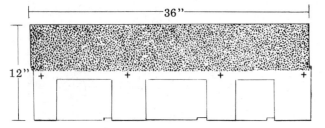

### BIRD ARCHITECT® 70 SHINGLE

Shingle specifications: 25 year guarantee, U/L type class C.
Approximate weight: 345 pounds per square.
Dimensions: Length 36 inches; width 12 inches.
Exposure: 5 inches; Headlap 2 inches.
Bundles per square: 5. Minimum pitch: 4 in 12
Shingles per square: 80
Nails per shingle: 4
1¼ in. nails per square: 2 lbs.; 1½ in. nails per square: 2¼ lbs.

Hip and Ridge shingles: Use Bird hip and ridge shingles.

Application procedure: Apply an 18 in. wide strip of mineral surface roll roofing at the bottom edge with a ¼ in. overhang. Cut 5 in. off of the butt edge of an Architect 70 shingle and use the remaining 7 inches for the starter course. The first piece of starter should have 6 inches cut from one end. Continue laying full length starter pieces across the bottom. The first shingle must be a whole shingle, aligned with the starter course. The second and succeeding shingles up the rake should be laid at random. However, the end joints of a given shingle should offset the end joints of the underlaying shingle at least 6 inches. Also, the end joint of a given shingle must be at least 3 inches from the nails of the underlaying shingle. The proper exposure can be maintained by using the alignment steps at the ends of the shingles. The shingles must be properly nailed so that the nails penetrate both layers of the shingle.

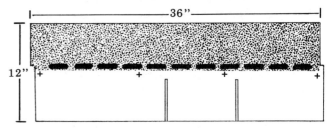

### BIRD ARCHITECT MARK 25® SHINGLE

Shingle specifications: 25-year guarantee, U/L type Class C and wind resistant.
Approximate weight: 290 lbs. per square.

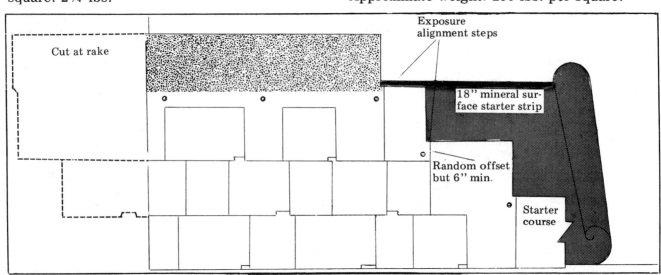

Bird Architect® 70 shingle
Starting procedure

Dimensions: Length 36 in.; width 12 in.
Exposure: 5 in.; Headlap 2 in.
Bundles per square: 4. Minimum pitch 4 in 12
Shingles per square: 81
Nails per shingle: 4
1¼ in. nails per square: 2 lbs.; 1½ in. nails per square: 2¼ lbs.
Hip and Ridge: Cut Mark 25 shingles into thirds or use hip and ridge shingles
Application procedure: The same as the Bird Architect 70 shingle.

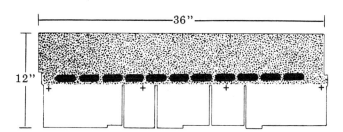

**BIRD WOODSMAN™ SHINGLE**
Shingle specifications: 20-year guarantee, U/L type class C and wind resistant
Approximate weight: 260 lbs. per square
Dimensions: Length 36 in.; width 12 in.
Exposure: 5 in.; Headlap 2 in.
Bundles per square: 4. Minimum pitch 4 in 12
Shingles per square: 81
Nails per shingle: 4
1¼ in. nails per square: 2 lbs.; 1½ in. nails per square: 2¼ lbs.
Hip and Ridge: Use hip and ridge shingles.
Application procedure: Starter course, cut evenly along the top of the cutouts and use the top portion for starter. Begin at the rake using a 30 inch piece of starter and continue along the bottom edge. The first shingle will be a full shingle aligned with the starter at the rake and eave. At the top of the Woodsman shingle are 5 alignment slits; one in the center and two on each side of the center slit. For a left handed corner start, you should cut the second shingle at the first slit from the left end. Position this remaining portion of shingle with the cut edge at the rake and align the shingle using the alignment steps. The third shingle will be cut at the second slit and nailed into place. The fourth shingle will be cut at the center slit. The fifth shingle will start over again with a full shingle. Repeat the same procedure as with the 2nd, 3rd and 4th shingles until the shingles are up to the top of the slope. The cut off portions can be used at the other end of the roof.

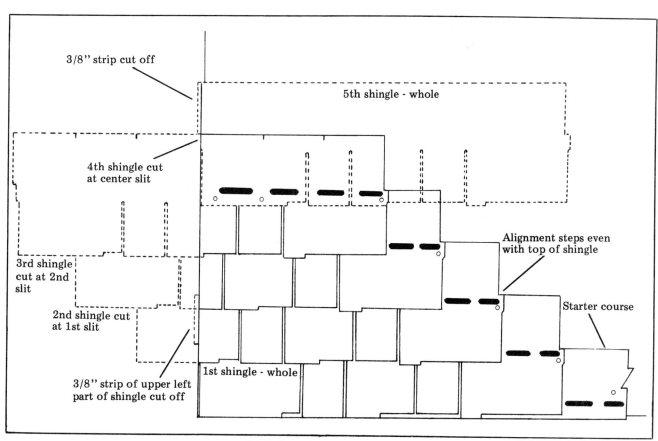

Bird Woodsman™ shingle
Starting procedure

Roofers Handbook

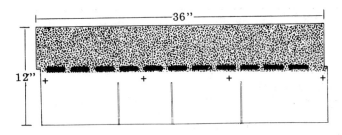

### BIRD WIND SEAL JET® SHINGLE

Shingle specifications: 15-year guarantee, mono-tab (no cut outs) U/L type class C and wind resistant
Approximate weight: 237 lbs. per square
Dimensions: Length 36 in. width 12 in.
Exposure: 5 in.; Headlap 2 in.
Bundles per square: 3
Shingles per square: 80
Nails per shingle: 4
1¼ in. nails per square: 2 lbs.; 1½ in. nails per square: 2¼ lbs.
Hip and Ridge: Cut Jet shingles into thirds or use standard hip and ridge units

Application procedure: Same as the Bird Architect 70 shingle.

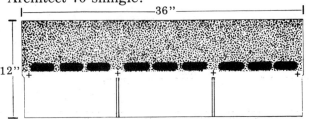

### BIRD WIND SEAL® SHINGLE

Shingle specifications: 15-year guarantee, 3 tab, U/L type class C and wind resistant
Approximate weight: 235 lbs. per square
Dimensions: Length 36 in.; width 12 in.
Exposure: 5 in.; Headlap: 2 in.
Bundles per square: 3
Shingles per square: 80
Nails per shingle: 4
1¼ in. nails per square: 2 lbs.; 1½ in. nails per square: 2¼ lbs.
Hip and Ridge: Cut the shingles into thirds or use hip and ridge shingles.
Application procedure: Standard 3-tab methods.

# Celotex® Roofing Materials

Celotex Building Products
The Celotex Corporation
A Jim Walter Company
P. O. Box 22602, Tampa, Florida 33622
Sales Offices:
1900 MacArthur Blvd. N.W., Atlanta, Georgia 30318
640 Pearson Street, Chicago, Illinois 60016
320 S. Wayne Ave., Cincinnati, Ohio 45215
13601 Preston Road, Dallas, Texas 75240
One River Road, Edgewater, New Jersey 07020
2210 West 75th Street, Shawnee Mission, Kansas 66208
1633 N. Pablo St., Los Angeles, California 90033
or
P. O. Box 31178, Lincoln Heights Station, Los Angeles, California 90031
36th and Grays Ferry Avenue, Philadelphia, Pennsylvania 19146
Note: Specifications current as of January 1976
Contact manufacturer for text of warranties

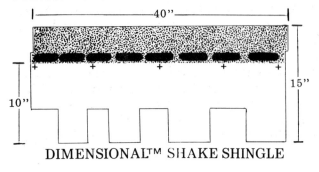

### DIMENSIONAL™ SHAKE SHINGLE

Shingle specifications: 25-year limited warranty, U/L type class C
Dimensions: Length 40 in.; width 15 in.
Exposure: 4 in. avg.; Headlap 6 in.
Bundles per square: 5
Shingles per square: 90
Nails per shingle: 5
1¼ in. nails per square: 2¾ lbs.; 1½ in. nails per square: 3 lbs.
Dimensional Shake Hip and Ridge Shingles: Size, 12 in. by 12 in., 4 bundles per 100 lineal feet, 60 shingles per bundle, (5 in. exposure).
Application procedure: Starter strip. At the eave apply an 18 in. wide strip of mineral surface roll roofing (any color), nail every 6 inches in a row 9 in. from the bottom edge. Over this strip apply the upper 10 inches of Dimensional Shake Shingles (cut off 5 in. tabs), using five nails per shingle 6 in. from the bottom edge. Start with a portion of shingle so that the joint between shingles in the starter course will be under the tab of the shingle in the first course. Cut two inches from first piece of starter and align it with the mineral surface roofing. Continue the starter course with full length shingles.

Applying shingles: Shingles may be started from the middle, left or right side of the roof. Start the first course with a full shingle allowing the 5 in. tabs to project off the bottom edge ½ inch. Cement these tabs down with daubs of

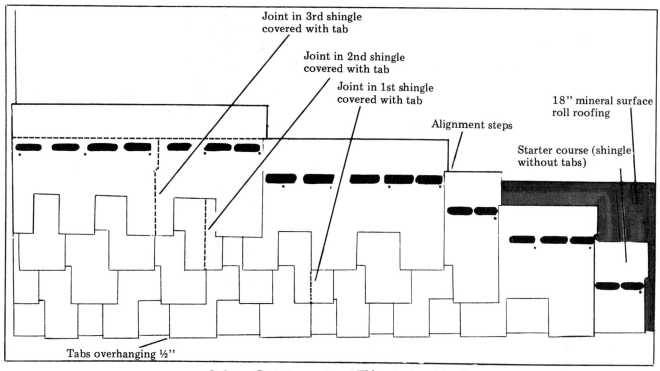

Celotex® Dimensional™ shake shingles
Starting procedure

plastic cement about the size of a quarter. Start the second course, offsetting either left or right, so that the first or second tab covers the joint in the shingle below. Continue subsequent courses, offsetting shingles either left or right, making certain that joints in the course below are always covered by a tab. Do not align similar sized tabs in either a vertical or diagonal line. Vary the offset in each course. However, the offset should not be less than 1½ inch. If a joint is exposed because of a change in shingle pattern, lay the shingle so that the joint is covered by a tab. Fill the gap with an insert not less than 6 inches wide. The insert can be cut from a scrap shingle. As each course is laid, align notches at the ends of the shingle to assure a 1 inch overhang of the tabs. Butt the ends snugly. Strike chalk lines every six courses to check parallel alignment with the eaves.

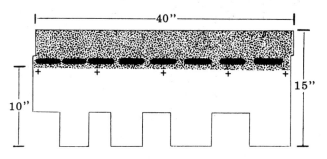

**CELOTEX® TRADITIONAL SHAKE SHINGLE**

Shingle specifications: Random tab self sealing, 25-year limited warranty, U/L type class C
Dimensions: Length 40 in.; width 15 in.
Exposure: 4 in.; avg. Headlap 6 in.
Bundles per square: 5
Shingles per square: 90
Nails per shingle: 5
1¼ in. nails per square: 2¾ lbs.; 1½ in. nails per square: 3 lbs.
Traditional Hip and Ridge Shingles: Size, 12 in. by 12 in., 4 bundles per 100 lineal feet, 60 shingles per bundle (5 inch exposure).
Application procedure: At eaves apply an 18" wide strip of mineral surface roll roofing. Nail every six inches in a row 9 inches from the bottom edge. Over this strip apply a 12" wide strip of mineral surface roll roofing the same color as the shingles. Nail every 6 inches in a row 6 inches from the bottom edge.
Applying shingles: Shingles may be applied from either the left or the right side of a roof. However, application should be made in only one direction to assure proper installation. Start the first course with a full shingle flushed with the eave. Seal down the tabs with plastic cement. Start the second course with a shingle cut at the left top notch for a left hand start and the right top notch for a right hand start. The self-alignment notch at the ends assures a 1 in. overhang of the tabs. Start the third course with a shingle cut at the right top notch for a left hand start or left top notch for a right hand start.

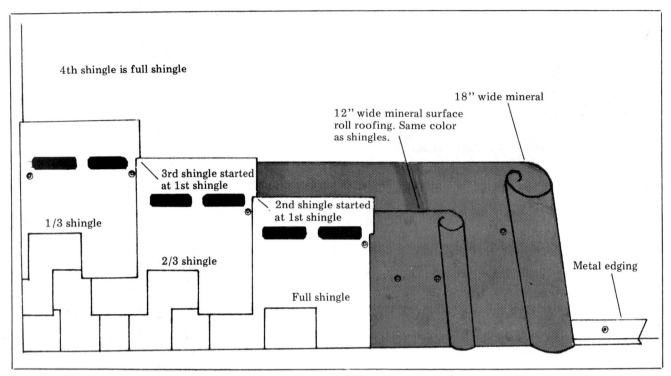

Celotex® Traditional™ shake shingles
Starting procedure

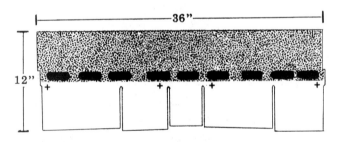

**CELOTEX® RUSTIC SHINGLE**
Shingle specifications: Random tab self-sealing, 18-year limited warranty, U/L type class C and wind resistant
Dimensions: Length 36 in.; width 12 in.
Exposure: 4¾ in.; Headlap 2¼ in.
Bundles per square: 3
Shingles per square: 84

Nails per shingle: 4
1¼ in. nails per square: 2 lbs.; 1½ in. nails per square: 2¼ lbs.
Rustic Hip and Ridge Shingles: 12 in. by 12 in. shingles, 3 bundles per 100 lineal feet. 80 shingles per bundle (5 inch exposure).
Application procedure: Same as the Rustic Shakes® Fire-Chex® Shingles.

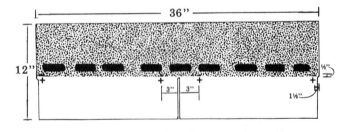

**CELOTEX® FIRE-CHEX® SHINGLE**
Shingle specifications: 2-tab, self-sealing, 25-year limited warranty, U/L type class A and wind resistant
Dimensions: Length 36 in.; width 12 in.
Exposure: 5 in.; Headlap 2 in.
Bundles per square: 4
Shingles per square: 80
Nails per shingle: 4

*Reference Section*

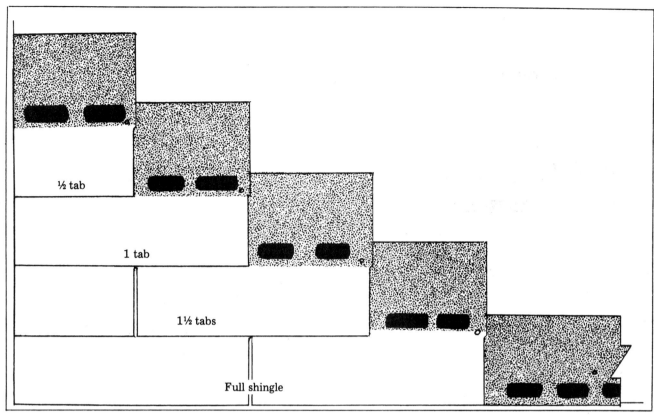

Celotex Fire-Chex® shingles
Starting procedure

1¼ in. nails per square: 2 lbs.; 1½ in. nails per square: 2¼ lbs.
Fire-Chex Hip and Ridge Shingles: Pre-formed shingles; 12 in. wide tab. 6 bundles per 100 lineal feet, 40 shingles per bundle. (5 in. exposure).
Application Procedure: For starter strip use the upper portion of Fire-Chex shingles with the tabs cut off even. Start at the rake with a piece 28 inches long. Continue across the roof with full length pieces. Start the first course with a full shingle, second course with 1½ tabs, third course with 1 tab and the fourth course with a ½ tab. The fifth course will start over again with a full shingle.

resistant
Dimensions: Length 36 in., width 12 in.
Exposure: 5 in.; Headlap 2 in.
Bundles per square: 3
Shingles per square: 80
Nails per shingle: 4
1¼ in. nails per square: 2 lbs.; 1½ in. nails per square: 2¼ lbs.
Hip and Ridge: No special ridge necessary. Use standard Hip and Ridge.
Application procedure: Application methods are similar to the standard 3-tab shingle. Lay at a random pattern but be sure that the end joints are offset as per manufacturer's recommendation on the bundle wrapper.

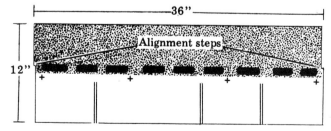

CELOTEX® STORM KING™ SHINGLE
Shingle specifications: No-cutout shingle, random embossed, self-sealing, mono-tab, 15-year limited warranty U/L type class C and wind

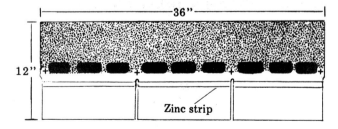

CELOTEX® FRS® SOL-SEAL/15 SHINGLE
AND FRS® REGENCY/25
Shingle specifications: FRS® (Fungus resistant shingle) 3-tab, self-sealing, white only, 15-year

limited warranty, U/L type class C and wind resistant
Dimensions: Length 36 in.; width 12 in.
Exposure: 5 in.; Headlap 2 in.
Bundles per square: 3
Shingles per square: 80
Nails per shingle: 4
1¼ in. nails per square: 2 lbs.; 1½ in. nails per square: 2¼ lbs.
Hip and Ridge: Cut the necessary ridge from FRS Sol-Seal/15 Shingles or FRS Regency/25.
Application Procedure: Standard 3-tab shingle application methods. Only exception is that the shingle should never have an exposure less than 5 inches. The reason is that the fungus repelling granules are applied to the shingle at the top of the tabs in a ½ inch band across the entire length of the shingle. A 4½ inch exposure anywhere on the roof would cover up this part of the shingle.

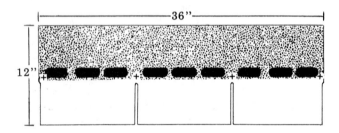

### CELOTEX® SOL-SEAL SHINGLE

Sol-Seal/15 Shingle; standard weight, 15-year limited warranty

Sol-Seal/20 Shingle; medium weight, 20-year limited warranty

Regency/25 Shingles; heavy weight, 25-year limited warranty

Shingle specifications: 3-tab, self-sealing, U/L type class C and wind resistant
Dimensions: Length 36 in.; width 12 in.
Exposure: 5 in.; Headlap 2 in.
Bundles per square: Sol-Seal/15 and Sol-Seal/20, (3), Regency/25, (4)
Shingles per square: 80
Nails per shingle: 4
1¼ in. nails per square: 2 lbs.; 1½ in. nails per square: 2¼ lbs.
Hip and Ridge: For the standard weight shingle, use either hip and ridge units or cut shingles into thirds. For the medium and heavy weight shingles, cut the ridge from the same shingles that are being applied. Medium ridge shingles for a medium roof and heavy ridge shingles for a heavy weight roof.
Application procedure: Standard 3-tab methods.

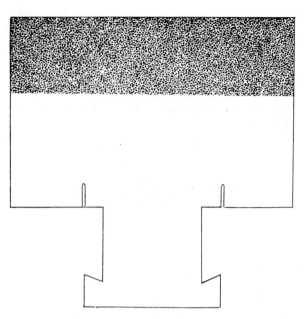

### DOUBLE-COVERAGE SAFE-LOX® SHINGLE

Shingle specifications: 15-year limited warranty, U/L type class C and wind resistant
Dimensions: 21.6'' x 22.22''
Exposure: 14-3/8'' x 11'' avg.; Headlap 4½''
Bundles per square: 3
Shingles per square: 90
Nails per shingle: 2
1¼ in. nails per square: 1¼ lbs.; 1½ in. nails per square: 1½ lbs.
Hip and Ridge: Use standard hip and ridge shingles.
Application procedure: Standard T-lock methods.

### LOKON SHINGLE

Shingle specifications: 15-year limited warranty, U/L type class C and wind resistant
Dimensions: 20'' x 20''
Headlap: 3½ in.
Bundles per square: 3
Shingles per square: 111
Nails per shingle: 2
1¼ nails per square: 1½ lbs.; 1½ in. nails per square: 1¾ lbs.
Hip and Ridge: Use the standard hip and ridge shingles.
Application procedure: Standard T-lock methods.

### STANDARD SAFE-LOX® SHINGLE

Shingle specifications: 15-year limited warranty, U/L type class C and wind resistant
Dimensions: 22½'' x 22.22''
Exposure: 18'' x 11'' avg. Headlap 2-1/8''
Bundles per square: 3
Shingles per square: 72
Nails per shingle: 2

1¼ in. nails per square: 1 lb.; 1½ in. nails per square: 1¼ lbs.
Hip and Ridge: Use standard hip and ridge shingles.
Application procedure: Standard T-lock methods.
Accessories:
Standard Hip and Ridge shingles: 9'' x 12'', 240 shingles (2 bundles) per 100 lineal feet with 5 in. exposure.
18 in. Mineral Surface Roll roofing: 36 ft. long with 54 sq. ft. to a roll.
9 in. Mineral Surface Roll roofing: 36 ft. long with 27 sq. ft. to a roll.

# Certain-teed Roofing Materials

Certain-teed Products Corporation
P. O. Box 860
Valley Forge, Pennsylvania 19482

Regional Sales Offices:
1 Regency Drive, Bloomfield, CT 06002
Box 13703, 6605 Abercorn St., Savannah, GA 31406
191 Joe Orr Rd., Chicago Heights, IL 60461
5700 Broadmoor, Suite 1002, Mission, KS 66202
5909 Orchard Street West, Tacoma, WA 98467
2502 Silverside Rd., Wilmington, DE 19810
Box 1700, Rt. 250 & Ohio Turnpike, Milan, OH 44846
3303 East 4th Avenue, Shakopee, MN 55379
Oak Cliff Bank Tower, Suite 815, 400 S. Zangs, Dallas, TX 75208
Watergate Tower, Suite 495, 1900 Powell St., Emeryville, CA 94608

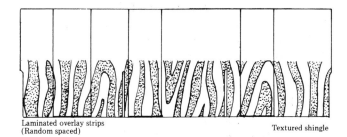
Laminated overlay strips (Random spaced)   Textured shingle

### HALLMARK SHANGLE®, TEXTURED
Shingle specifications: 25-year warranty, U/L type class C
Approximate weight: 380 lbs. per square
Dimensions: Length 36 in.; width 12¼ in.
Exposure: 5-1/8 in.; Headlap 2 in.
Bundles per square: 5
Shingles per square: 78
Nails per shingle: 4
1¼ in. nails per square: 2 lbs.; 1½ in. nails per square: 2¼ lbs.
Hip and Ridge shingles: Cut the ridge shingles from Certain-teed Woodtex Shingles of the same color. For a distinctive appearance, lay the shingles in a double layer.

Application-Hallmark
Starter strip: Install an 18'' wide Mineral Surfaced Starter Strip overhanging the eaves and rakes ½'' to ¾''.

Starter Course: (a) Preparation: The Starter Course consists of the upper 7-1/8'' portion of Hallmark Shingles with the overlay strips left intact. Starter Course shingles are fabricated by simply cutting off the 5-1/8'' tabs from the shingle slabs. (b) Application: Cut 6'' off the length of the first slab of 7-1/8'' wide Starter Course and apply over the Starter Strip beginning at the lower left hand corner of the roof. This first 30'' long Starter Course strip shall be flush with the Starter Strip ½'' to ¾'' at the eaves. Succeeding Starter Course slabs shall be full length (36''). This will provide a drip-edge which will project 1'' to 1½'' beyond the eaves.

Shingle Application: (a) Important: Shingles must be applied using a diagonal application method. *Never* apply shingles in continuous horizontal or vertical courses. (b) It is necessary to begin roofing at the left rake-edge of the roof starting each shingle course beginning at the eave with a single slab cut to a special length. The first shingle of every course must be cut to a special length to assure the succeeding shingles applied full-length in the field of the roof develops the desired appearance.

*Step 1.* Starting at the left rake edge and the eave, apply one full length (36'') shingle flush with the starter course. This will be the first course starter shingle.

*Step 2.* Next, for the second course starter shingle, cut 4'' off the left tab of a shingle and apply the 32'' long section over the headlap of the first course shingle, exposing the first course 5-1/8''.

*Step 3.* The starter course shingles for courses 3 through 9 are to be cut to exact dimensions and installed in sequence as the next steps in application. Cuts are to be made from the left side of each slab.

| Courses | Cut From Left Tab | Length Of Shingle To Be Applied |
|---|---|---|
| 3 | 8″ | 28″ |
| 4 | 12″ | 24″ |
| 5 | 16″ | 20″ |
| 6 | 20″ | 16″ |
| 7 | 24″ | 12″ |
| 8 | 28″ | 8″ |
| 9 | 32″ | 4″ |

Up to this point the first shingle of the first nine courses has been applied from the eave up the rake and a diagonal application pattern has been established. Beginning with the full length (36″) first course starter shingle, succeeding starter shingles thus installed at the rake are 4″ shorter progressively.

*Step 4.* Return to the starting point and apply a full length (36″) shingle next to the bottom starter shingle. Continue diagonally up the roof applying full length (36″) shingles through the ninth course. Continue applying full length (36″) shingles in a diagonal sequence from course 1 through 9 as far across the roof as desired.

*Step 5.* The 10th through 18th courses should be applied in exactly the same manner as the 1st through 9th. If more shingle courses are required above 18, repeat the same procedure. Note: Pieces cut from the shingles used along the left rake can be used to finish off the courses at the right rake.

High Wind Areas - When Hallmark, Independence and Woodtex are used in areas where the roof is exposed to excessively strong winds, 6 nails per shingle are recommended. They are to be located 5¾″ up from the lower edge of the tabs, 1″ in from each end of the shingle and 1½″ to either side of the cutouts. It is also advisable to cement the tabs of the shingles.

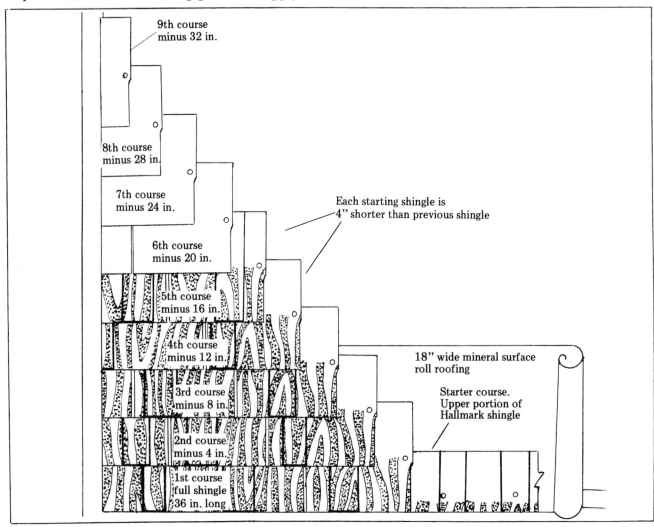

Certain-teed Hallmark Shangle®
Starting procedure

This is easily done by placing a spot of plastic cement or sealing cement, the size of a half dollar, directly below the lower center of the tab. Press tab firmly in place. The spot of cement should be so placed that when the tab of the shingle is pressed down the cement is squeezed just to the edge of the tab.

Note: Closed valley construction (half lace) is recommended. Remove the overlay strips that extend past and across the valley. Valleys should be lined with 36" wide roll roofing either smooth or mineral surfaced. Extend shingles of the underlying side at least 12 inches beyond the valley center. Shingles of the overlying side should be trimmed in a straight line 1" from the center at the top and 2" at the bottom.

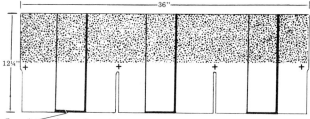
Laminated overlay strips

### INDEPENDENCE SHANGLE™

Shingle specifications: 25-year warranty, U/L type class C
Approximate weight: 340 lbs per square
Dimensions: Length 36 in.; width 12¼ in.
Exposure: 5-1/8" in.; Headlap 2 in.
Bundles per square: 4
Shingles per square: 78
Nails per shingle: 4
1¼ in. nails per square: 2 lbs.; 1½ in. nails per square: 2¼ lbs.
Hip and Ridge shingles: Independence Accessory and Hip and Ridge shingles 12¼" x 36", 25 shingles (1 bundle) per 16 lineal feet (double layers) with a 5-1/8" exposure. Approximate weight per bundle: 86 lbs.
Application procedure: Same as the Hallmark Shangle®

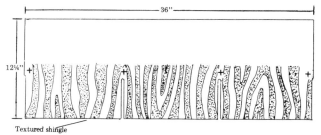
Textured shingle

### WOODTEX® SHINGLE

Shingle specifications: 25-year warranty, U/L type class C.
Approximate weight: 300 lbs. per square
Dimensions: Length 36 in., width 12¼ in.
Exposure: 5-1/8 in.; Headlap 2 in.
Bundles per square: 4
Shingles per square: 78
Nails per shingle: 4
1¼ in. nails per square: 2 lbs., 1½ in. nails per square: 2¼ lbs.
Hip and Ridge shingles: Cut from Woodtex® shingle or Woodtex® Hip and Ridge.
Application procedure: Application methods are the same as the standard 3-tab shingle.

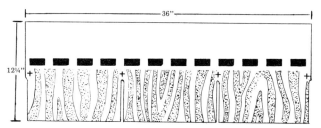

### GLASSTEX® SHINGLE

Shingle specifications: 20-year warranty, U/L type class A, and UL Wind Resistance Rated.
Approximate weight: 260 lbs. per square
Dimensions: Length 36 in.; width 12¼ in.
Exposure: 5-1/8 in.; Headlap 2 in.
Bundles per square: 4
Shingles per square: 78
Nails per shingle: 4
1¼ in. nails per square: 2 lbs.; 1½ in. nails per square: 2¼ lbs.
Hip and Ridge shingles: Cut ridge from the same Glasstex shingles.
Application procedure: Application methods are the same as the standard 3-tab shingle.

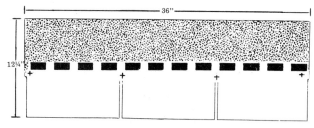

### GLASSGUARD® SHINGLE

Shingle specifications: 20-year warranty, U/L type class A, and UL Wind Resistance Rated.
Approximate weight: 215 lbs. per square
Dimensions: Length 36 in.; width 12¼ in.
Exposure: 5-1/8 in.; Headlap 2 in.
Bundles per square: 3
Shingles per square: 78
Nails per shingle: 4
1¼ in. nails per square: 2 lbs.; 1½ in. nails per square: 2¼ lbs.
Hip and Ridge shingles: Cut from the same Glassguard shingles.

Application procedure: Application methods are the same as the standard 3-tab shingle.

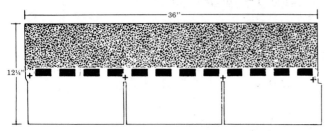

### STRIP SEAL SHINGLE

Shingle specifications: No cut-out (available at Shakopee plant only), 15-year warranty, U/L type class C and wind resistant.
Approximate weight: 235 lbs. per square
Dimensions: Length 36 in.; width 12¼ in.
Exposure: 5-1/8 in.; Headlap 2 in.
Bundles per square: 3
Shingles per square: 78
Nails per shingle: 4
1¼ in. nails per square: 2 lbs.; 1½ in. nails per square: 2¼ lbs.
Hip and Ridge shingles: Use standard hip and ridge or cut from the same Strip Seal shingles or Sealdon® shingles.
Application procedure: Application methods are similar to the standard 3-tab shingle. Lay at random pattern but be sure that the end joints are offset per manufacturer's recommendations on the bundle wrapper.

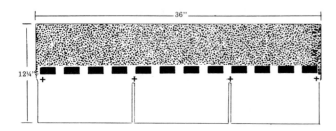

### CLASSIC SEALDON™ SHINGLE

Shingle specifications: 25-year warranty, U/L type class C and wind resistant
Approximate weight: 300 lbs. per square
Dimensions: Length 36 in.; width 12¼ in.
Exposure: 5-1/8th; Headlap 2 in.
Bundles per square: 4
Shingles per square: 78
Nails per shingle: 4
1¼ in. nails per square: 2 lbs.; 1½ in. nails per square: 2¼ lbs.
Hip and Ridge shingles: Cut from the same Classic Sealdon shingles.
Application procedure: Application methods are the same as the standard 3-tab shingle.

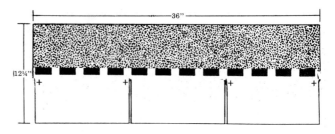

### CUSTOM SEALDON™ SHINGLE

Shingle specifications: 20-year warranty (available at the York, Pa. plant only), U/L type class C and wind resistant
Approximate weight: 265 lbs. per square
Dimensions: Length 36 in.; width 12¼ in.
Exposure: 5-1/8th in.; Headlap 2 in.
Bundles per square: 3
Shingles per square: 78
Nails per shingle: 4
1¼ in. nails per square: 2 lbs.; 1½ in. nails per square: 2¼ lbs.
Hip and Ridge shingles: Cut from the same Custom Sealdon shingles.
Application procedure: Application methods are the same as the standard 3-tab shingle.

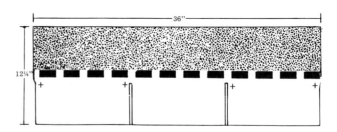

### SEALDON™ SHINGLE

Shingle specifications: 15-year warranty, UL type Class C and wind resistant, available at all plants. Also available at Dallas and Savannah only, Perma-Brite (star white color) with fungus growth retardation properties.
Approximate weight: 235 lbs. per square
Dimensions: Length 36 in.; width 12¼ in.
Exposure: 5-1/8th in.; Headlap 2 in.
Bundles per square: 3
Shingles per square: 78
Nails per shingle: 4
1¼ in. nails per square: 2 lbs.; 1½ in. nails per square: 2¼ lbs.
Hip and Ridge shingles: Cut from standard Sealdon shingles or use the standard hip and ridge shingles. NOTE: For the Perma-Brite shingles, cut ridge from the same fungus resistant shingles.
Application procedure: Standard 3-tab methods.

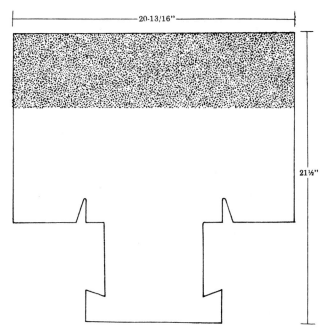

**CUSTOM SAF-T-LOK SHINGLE**
Shingle specifications: 17-year warranty, U/L type class C and wind resistant
Approximate weight: 250 lbs. per square
Dimensions: Length 21½ in.; width 20-13/16 in.

Exposure: 14 in.; Headlap 7 in.
Bundles per square: 3
Shingles per square: 99
Nails per shingle: 2
1¼ in. nails per square: 1¼ lbs.; 1½ in. nails per square: 1½ lbs.
Hip and ridge shingles: Use standard hip and ridge shingles.

Certain-teed accessories
Standard hip and ridge:
Richmond, Los Angeles, Tacoma and Portland facilities - 9" x 12", 240 shingles (3 bundles) per 100 lineal feet with 5 in. exposure. Approximate weight 180 lbs. Dallas and Savannah plants - 9" x 12", 240 shingles (2 bundles) per 100 lineal feet, with 5 in. exposure. Approximate weight 150 lbs. All other plants except Shakopee and York - 9" x 12", 240 shingles (2 bundles) per 100 lineal feet with 5 in. exposure. Approximate weight 170 lbs. Starter rolls (mineral surface roll roofing). Length 36 ft., width 9 in., 27 sq. ft., 22 22 lbs. per roll

Starter and valley. Length 36 ft., width 18 in. 54 sq. ft., 45 lbs.

# Flintkote Roofing Materials

The Flintkote Company
Building Materials Division
480 Central Avenue
East Rutherford, New Jersey 07073

Approximate weight: 390 lbs. per square
Dimensions: Length 40 in.; width 11½ in.
Average exposure: 4½ in.; Headlap 2 in.
Bundles per square: 5
Shingles per square: 80
Nails per shingle: 4
1½ in. nails per square, 2¼ lbs.

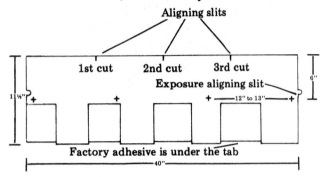

SIERRA® '390'
Shingle specifications: 25-year guarantee, U/L type class C and wind resistant
A self sealing shingle with staggered butts, irregular in width and thickness. Alignment slits at the ends and top of the shingle.

Application procedure: A definite starting procedure must be used to insure the proper pattern that has been designed into the shingle. Note the diagram. The starter is nine inch mineral surface roll roofing. The first shingle is a whole shingle with the bottom edge of the underlay portion even with the bottom edge of the starter. The second shingle laid must be cut at the first aligning slit from the left end of the shingle. Align the cut edge even with the edge of the whole shingle. The third shingle must be cut at the center aligning slit and the right portion slid over and nailed into place. The fourth shingle must be cut at the slit next to the

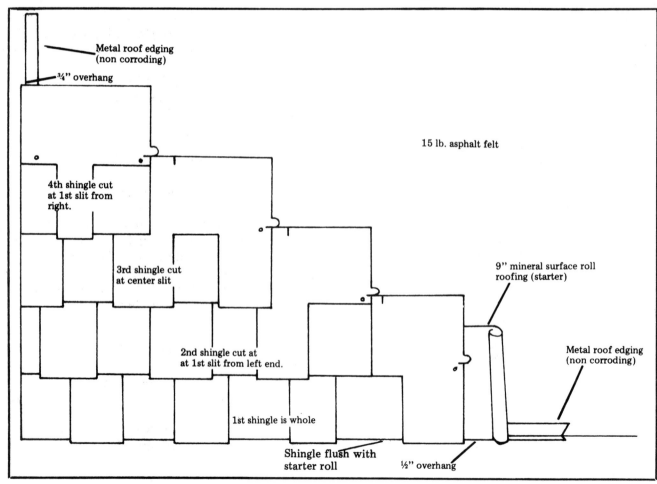

Flintkote® Sierra® '390'
Starting procedure

*Reference Section*

right end and the smaller portion can then be nailed into place. The fifth shingle will be a whole shingle after a row of shingles has been brought up from the bottom. The remaining portions of the shingles can be used at the other end of the slope. The sixth, seventh and eight courses will be a repeat of the second, third and fourth shingles.

The nails should be 11 or 12 gauge with a 3/8 or 7/16 in. head, 1½ in. long. The proper nailing area is six inches below the top of the shingle, one inch in from each end. The remaining two nails should be spaced approximately 12 to 13 inches apart and 12 to 13 inches from the end nails.

Ridge: Use the twin hip and ridge of a similar color. Apply in a double layer.

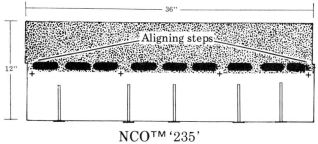

NCO™ '235'

Shingle specifications: Mono-tab, no cut outs, U/L type class C and wind resistant
Approximate weight: 235 lbs.
Dimensions: Length 36 in.; width 12 in.
Exposure: 5 in.; Headlap 2 in.
Bundles per square: 3
Shingles per square: 80
Nails per shingle: 4
1¼ in. nails per square: 2 lbs.; 1½ in. nails per square: 2¼ lbs.
Application procedure: Shingles are applied similar to the three tab strip shingle. The shingles are random embossed and have a smooth, no pattern appearance. Each shingle must offset the shingle under it about six inches. Be careful that a joint does not fall directly over a nail in the underlaying shingles.

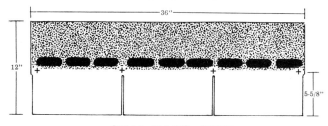

SEAL-TAB® '235'

Shingle specifications: U/L type class C and wind resistant
Approximate weight: 235 lbs. per square
Dimensions: Length 36 in.; width 12 in.
Exposure: 5 in.; Headlap 2 in.
Bundles per square: 3
Shingles per square: 80
Nails per shingle: 4
1¼ in. nails per square: 2 lbs.; 1½ in. nails per square: 2¼ lbs.
Standard three tab application methods.

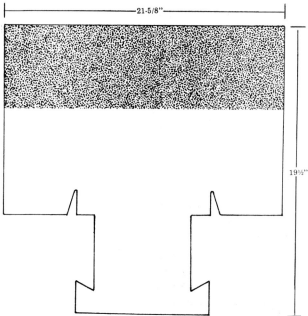

DOUBLE COVERAGE SHURLOK® '250'

Shingle specifications: U/L type class C and wind resistant
Approximate weight: 250 lbs. per square
Dimensions: 19½ by 21-5/8 inches
Bundles per square: 4
Shingles per square: 104
Nails per shingle: 2
1¼ in. nails per square: 1¼ lbs.; 1½ in. nails per square: 1½ lbs.
Hip and Ridge: Use standard hip and ridge or 3-tab shingles.
Asphalt shingle accessories: Twin Hip and Ridge '180', 2 colors on each shingle, U/L type class C, Size: 9 in. by 12 in., Weight: 180 lbs. per square, Bundles per square: 2, Shingles per square: 240, Exposure: 5 inches. One square of ridge will cover 100 lineal feet.

Valley liner:
Mineral surface roofing, No. 90 cut in strips for lining valleys.
Each roll is 18 in. wide and 36 ft. long.

Starter rolls:
Mineral surface roofing No. 90 cut in strips for starter courses beneath shingles. 9 in. by 36 ft. and 7-1/5 in. by 36 ft.
Additional shingles not described here are the Tripltab® Hexagon '195' and the Standard Shurlok® '180'.

# Fry Roofing Materials

Lloyd A. Fry Roofing Company
General Offices: 5818 Archer Rd.
Summit, Illinois 60501

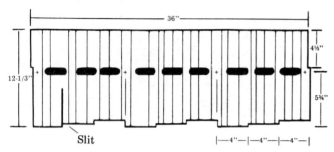

### BONDED SUBURBAN THATCH SEAL RITE SHINGLE

Shingle specifications: Bonded 20 years, U/L type class C and wind resistant
Approximate weight: 290 lbs. per square
Dimensions: Length 36 in.; width 12⅓ in.
Exposure: 4½ in. average; Headlap 2½ in. average
Bundles per square: 4
Shingles per square: 100
Nails per shingle: 4
1¼ in. nails per square: 2½ lbs. 1½ in. nails per square: 2¾ lbs.
Ridge: Suburban Thatch Shingles cut up in 12 in. widths are to be used.
Expose 5 inches. Do not use any slit pieces.

Application procedure: The starter course can be a Suburban Thatch Shingle with the bottom 4½ inches cut off straight. Start out with a half shingle (18 in. piece) in the lower left hand corner and allow a 3/8 in. overhang at eave and rake. Then continue with trimmed off shingles along the eave, overlapping each piece back onto the last starter shingle 4 inches. (Caution: Do not lay the starter shingles with the ends butted against each other as is done with other starter shingles as this may result in a leak).

First course: From a whole shingle cut off 4 inches (at slit) and lay the remaining 32 inches in the left hand corner directly over the starter shingle. The second course will have 12 inches cut off (8 inches more than the first shingle). The third course will have an additional 8 inches cut off (a total of 20 inches) leaving a 16 inch shingle to be nailed in place. The fourth course in the step in process will be only 8 inches long (a total of 28 inches cut off).

Now start back at the bottom laying whole shingles. The shingles must *not* be end butted as with most shingles. Each shingle must lap back on the adjoining shingle 4 inches. This is easy to do. Simply slide the shingle over until the slit butts against the other shingle. Align the other end with the aligning step at the top of the underlying shingle. Shingle up to the 8 inch

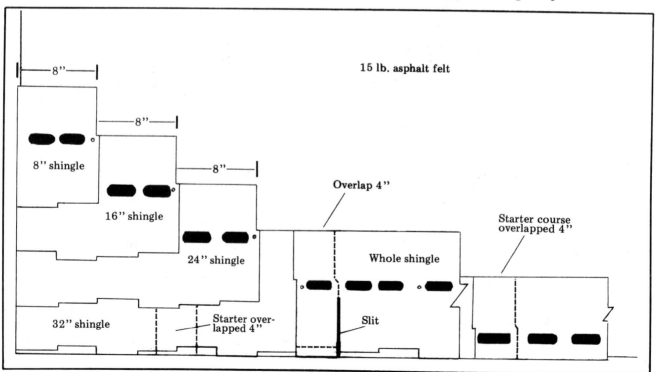

Fry Bonded Suburban Thach shingle
Application procedure

*Reference Section*

shingle, overlapping each shingle 4 inches. Now start the step-in process all over again by cutting 4 inches from the left end of a whole shingle. Now continue the same as with the second course, the third course, and so on.

The remaining portions of the shingles can be used at the other end of the roof slope. Time can be saved by precutting the four shingles and laying them out on the roof where you will be using them. The overlapping tabs should be sealed down with plastic cement (a spot the size of a quarter), particularly in high wind areas.

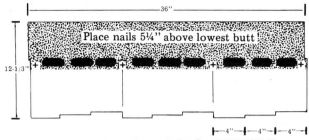

RUSTIC CRESTWOOD® SHINGLE
Shingle specifications: Bonded 20 years, U/L type class C and wind resistant
Approximate weight: 260 lbs. per square

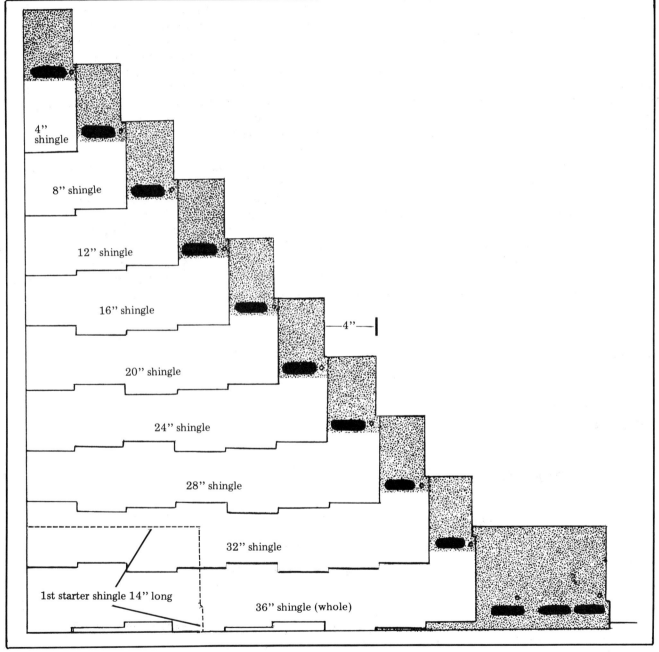

Fry Rustic Crestwood® shingle
Application procedure

Dimensions: Length 36 in.; width 12-1/3 in.
Exposure: 4½ in. average; Headlap 2½ in. average.
Bundles per square: 4
Shingles per square: 89
Nails per shingle: 4
1¼ in. nails per square: 2 lbs.; 1½ in. nails per square: 2¼ lbs.
Hip and Ridge: Cut from Rustic Crestwood® shingles, three 12 in. pieces.

Application procedure: Starter shingles can be cut from Rustic Crestwood® shingles. Trim off the lower 4½ inches and install with the factory adhesive at the bottom. Allow 3/8 in. to ½ in. overhang at the bottom edge and at the rake. Start with a 14 in. piece of starter.

First shingle: Start the first course with a whole shingle (minus the 3/8 in. ear at the upper left corner). Line the shingle up with the starter at the bottom edge and the rake. The second course must have 4 inches cut evenly from the left edge before application. The third course will have an additional 4 inches (total of 8 in.) cut from it. The fourth course will have an additional 4 in. (total of 12 in.) cut from it. Continue cutting an additional 4 inches off of the shingles until the last course is 4 inches wide (the ninth shingle). The tenth course will start over again with a whole shingle, then repeat the previous procedure.

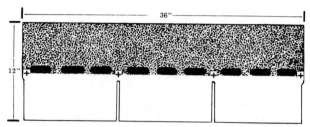

## BONDED SUPER 3-TAB SEAL RITE SHINGLE

Shingle specifications: Bonded 20-years, U/L type class C and wind resistant
Approximate weight: 255 lbs. per square
Dimensions: Length 36 in.; width 12 in.
Exposure: 5 in.; Headlap 2 in.
Bundles per square: 4
Shingles per square: 80
Nails per shingle: 4
1¼ in. nails per square: 2 lbs.; 1½ in. nails per square: 2¼ lbs.
Hip and Ridge: Cut from Bonded Super 3-tab shingles.
Application procedure: Standard 3-tab shingle application methods.

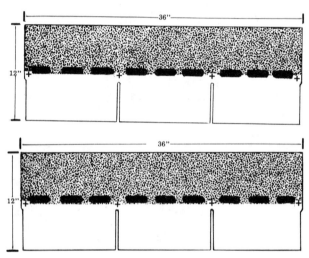

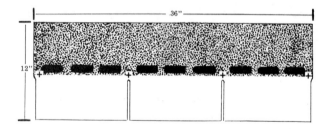

## BONDED GIANT THREE TAB SEAL RITE SHINGLE

Shingle specifications: Bonded 25-years, U/L type class C and wind resistant
Approximate weight: 290 lbs. per square
Dimensions: Length 36 in.; width 12 in.
Exposure: 5 in.; Headlap 2 in.
Bundles per square: 4
Shingles per square: 80
Nails per shingle: 4
1¼ in. nails per square: 2 lbs.; 1½ in. nails per square: 2¼ lbs.
Hip and Ridge: Cut from giant 3-tab shingles
Application procedure: **Standard 3-tab shingle application methods.**

## BONDED SPECIAL SEAL-RITE 3-TAB SHINGLE
and
STANDARD 3-TAB SEAL RITE SHINGLE

(Available in different colors than the Bonded Special Seal Rite Shingle).
Specifications for both shingles: Bonded 15 years (Bonded Special Seal Rite only), U/L type class C and wind resistant
Approximate weight: 235 lbs. per square
Dimensions: Length 36 in.; width 12 in.
Exposure: 5 in.; Headlap 2 in.
Bundles per square: 3
Shingles per square: 80
Nails per shingle: 4
1¼ in. nails per square: 2 lbs.; 1½ in. nails per square: 2¼ lbs.

Hip and Ridge: Cut from the same shingles
Application procedures: Standard 3-tab application methods.

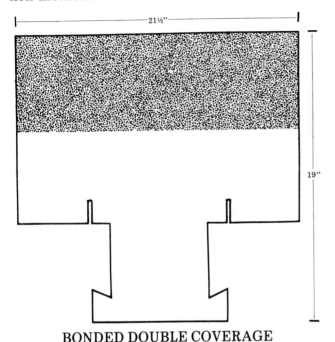

BONDED DOUBLE COVERAGE
LOCK SHINGLE
and
STANDARD DOUBLE COVERAGE
LOCK SHINGLE

Available in different colors than the Bonded D/C Lock Shingles.
Specifications for both shingles: Bonded 15-years, U/L type class C.
Approximate weight: 250 lbs. per square.

Dimensions: Length 19 in. width 21½ in.
Bundles per square: 3
Shingles per square: 108
Nails per shingle: 2
1¼ in. nails per square: 1½ lbs.; 1½ in. nails per square 1¾ lbs.

BONDED INDIVIDUAL SHINGLES (hip and ridge) Available in bonded shingle colors

STANDARD INDIVIDUAL SHINGLES (hip and ridge) Available in standard colors
Specifications for both: U/L type class C
Dimensions: Length 12 in.; width 9 in.
Shingles per square: 380
Bundles per square: 4
Approximate weight: 295 lbs. per square
126⅔ lineal feet coverage per square with a five inch exposure

SPECIAL MINERAL SURFACE
ROOF ROOFING (Available in bonded colors)
STANDARD MINERAL SURFACE
ROLL ROOFING (Available in standard colors)
U/L type class C, 108 sq. ft. per roll, 2 in. lap
Width: 36 in. Length 36 ft. Approximate weight: 90 lbs.

SPECIAL VALLEYS ROLLS
(Available in bonded colors)
STANDARD VALLEYS ROLLS
(Available in standard colors)
Width: 18 in., Length: 36 ft. Approximate weight per roll: 44 lbs.

# GAF Roofing Materials

GAF Corporation, Building Products
140 West 51st Street
New York, New York 10020

District Sales Offices:
P. O. Box 5166, Highlandtown, Baltimore, Maryland 21224
P. O. Box 5607, Dallas, Texas 75222
156 W. 56th Avenue, Denver, Colorado 80216
P. O. Box 1128, Erie, Pennsylvania 16512
P. O. Box 798, N. Broadway, Joliet, Illinois 60434
7600 Truman Rd., Kansas City, Missouri 64126
60 Curve Street, Millis, Massachusetts 02054
50 Lowry Ave. N., Minneapolis Minnesota 55411
P. O. Box 6377, Mobile, Alabama 36606
Givens Road, Mt. Vernon, Indiana 47620
P. O. Box 7329, Savannah, Georgia 31408

35 Main St., S. Bound Brook, New Jersey 08880
P. O. Box 5176, Tampa, Florida 33605

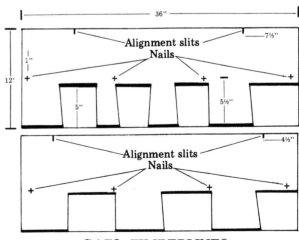

GAF® TIMBERLINE®
SELF-SEALING® SHINGLE

# Roofers Handbook

Shingle specifications: 25-year warranty, U/L class C, wind resistant label.
Approximate weight: 330 lbs. per square
Dimensions: Length 36 in.; width 12 in.
Exposure: 5 in.; Headlap 2 in.
Bundles per square: 4
Shingles per square: 80
Nails per shingle: 4
1¼ in. nails per square: 2 lbs.; 1½ in. nails per square: 2¼ lbs.
Hip and Ridge: Use GAF'S standard hip and ridge
Application procedure: Use 12 in. wide No. 90 mineral surface roll roofing for starter course and nail securely. Cut the first piece of starter about 7½ ft. long so that the joints will not coincide with the joints in the first course of shingles. Any additional starter pieces should be cut in 6 ft. or 9 ft. lengths.

First shingle will be a whole shingle aligned with the starter. Second shingle will be laid with the bottom edge even with the top of the saw tooth of the underlying shingle to obtain the 5 in. exposure. To offset it correctly, the right end should be set at the first alignment slit from the right end of the underlying shingle (4½ in. or 7½ in.). Third shingle applies the same way with the right end on the first slit from the right. If the second shingle was set on the 4½ in. slit, the third shingle should be set on the 7½ in. slit or vice versa. This is as far as the step-in goes. The fourth shingle will start over again with a whole shingle. The fifth and sixth shingles will be a repeat of the second and third shingles but they will not need to be in the same order. If the second shingle was set on the 4½ in. slit, then the fifth shingle can be set on the 7½ in. slit.

Strike a chalk line about every six courses to keep the shingles straight across the roof. Shingles can be laid from either the right side or the left side of the roof. When laying shingles from the left side, as in the drawing, use the alignment slits farthest from the rake edge. When laying shingles from the right side, use the alignment slit *closest* to the right rake edge. Hip roof application: Strike a vertical chalk line from the hip and ridge junction at the top to the eave to use as a guide for starting 1st, 4th, 7th etc. courses.

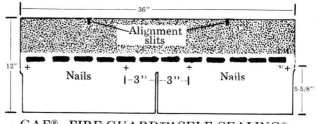

GAF® FIRE GUARD™ SELF-SEALING® SHINGLE

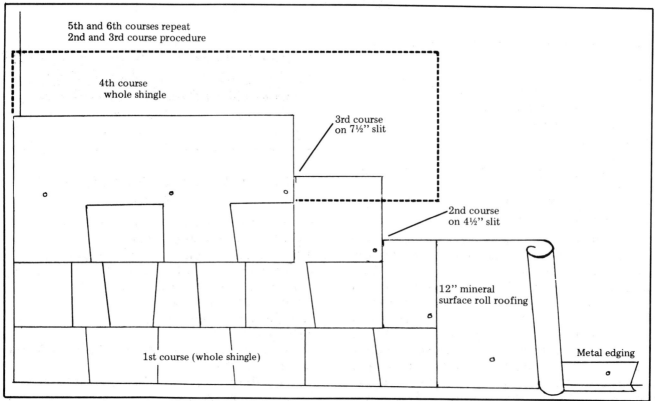

GAF Corporation Timberline® shingle
Starting procedure

*Reference Section*

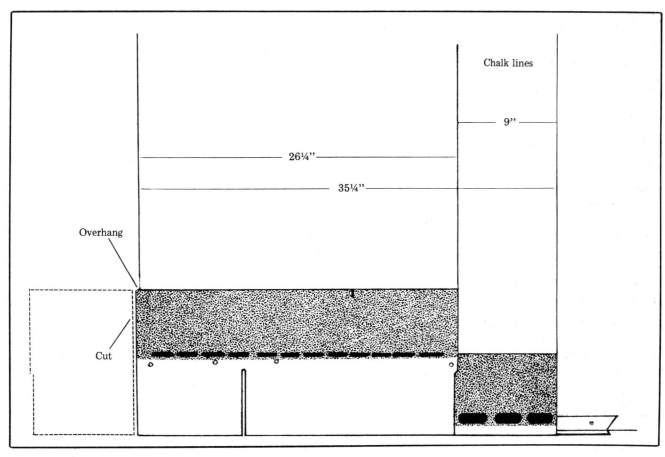

GAF® Fire Guard™ Self-Sealing™ shingle
Starting procedure

Shingle specifications: 25-year warranty, U/L class A, wind resistant label.
Approximate weight: 325 lbs. per square
Dimensions: Length 36 in.; width 12 in.
Exposure: 5 in.; Headlap 2 in.
Bundles per square: 4
Shingles per square: 80
Nails per shingle: 4
1¼ in. nails per square: 2 lbs.; 1½ in. nails per square: 2¼ lbs.
Hip and Ridge: From the same shingles cut 4 pieces of ridge, 9 in. by 12 in.
Application procedure: Application methods are similar to the standard 3-tab shingle, except the pattern should always be a half pattern. The starter course will be Fire Guard shingles with the tabs cut off even and the factory adhesive portion at the bottom edge. Two vertical chalk lines will help keep the bondlines straight up and down. From the rake edge measure in 35¼ in. and 26¼ in. at the top and bottom. Chalk the lines and start shingling with a whole shingle first, on the 35¼ in. line. The next shingle will be on the 26¼ in. line. The third shingle will be back on the 35¼ in. line, and so on.

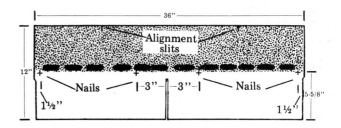

GAF® SOVEREIGN® SELF-SEALING™ SHINGLE

Shingle specifications: 25-year warranty, U/L class C, wind resistant label.
Approximate weight: 300 lbs. per square
Dimensions: Length 36 in.; width 12 in.
Exposure: 5 in.; Headlap: 2 in.
Bundles per square: 4
Shingles per square: 80
Nails per shingle: 4
1¼ in. nails per square: 2 lbs.; 1½ in. nails per square: 2¼ lbs.
Hip and Ridge: Cut from the same shingles four 9 in. by 12 in. pieces.
Application procedure: Same as the GAF Fire Guard Shingle.

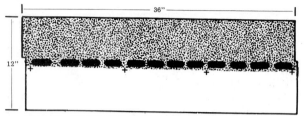

### GAF® NOR'EASTER™ SELF-SEALING™ SHINGLE

Shingle specifications: U/L class C, wind resistant label. Available in Millis, Mass. only.
Approximate weight: 235 lbs. per square
Dimensions: Length 36 in.; width 12 in.
Exposure: 5 in.; Headlap 2 in.
Bundles per square: 3
Shingles per square: 80
Nails per shingle: 4
1¼ in. nails per square: 2 lbs.; 1½ in. nails per square: 2¼ lbs.
Hip and ridge: Standard hip and ridge if available. If not then cut the ridge from the same shingle.
Application procedure: The shingles are applied similar to the standard 3-tab shingle, except without any vertical pattern. Each shingle is offset at random but with a six inch minimum. Care must be used to prevent joints from falling directly over a nail in the underlying shingle.

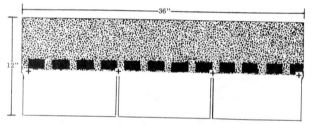

### GAF® STANDARD SELF-SEALING™ ASPHALT SHINGLE

Shingle specifications: U/L class C, wind resistant label.
Approximate weight: 235 lbs. per square
Dimensions: Length 36 in.; width 12 in.
Exposure: 5 in.; Headlap 2 in.
Bundles per square: 3
Shingles per square: 80
Nails per square: 4
1¼ in. nails per square: 2 lbs.; 1½ in. nails per square: 2¼ lbs.
Hip and ridge is available in the same colors or cut from 3-tab shingles.
Application procedure is the standard application for 3-tab shingles.

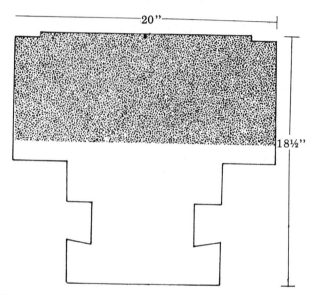

### GAF® TITE-ON® DOUBLE COVERAGE ASPHALT SHINGLE

Shingle specifications: U/L type class C and wind resistant
Approximate weight: 245 lbs. per square
Dimensions: Length 18½ in.; width 20 in.
Bundles per square: 3
Shingles per square: 120
Nails per shingle: 2
1¼ in. nails per square: 1¾ lbs.; 1½ in. nails per square: 2 lbs.
Hip and ridge is available in the same colors or cut from 3-tab shingles.

# Johns-Manville Roofing Materials

Johns-Manville Sales Corporation
Greenwood Plaza
Denver, Colorado 80217

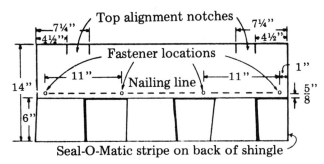

### THE WOODLANDS™ ROOF
(Fiber Glass Shingle)

Shingle specifications: U/L type class A and wind resistant.
Approximate weight: 300 lbs. per square.
Dimensions: Length 36¼ in.; width 14 in., except Manville, N. J. plant.
Exposure: 6 in.; Headlap 2 in. min.
Bundles per square: 4
Shingles per square: 67
Nails per shingle: 4
1½ in. nails per square: 2 lbs. 1¾ in. nails per square: 2¼ lbs.
Hip and Ridge Shingles: One bundle, shingle size 9 in. x 12 in., covers 25 lineal ft. when applied with double thickness.
Manville, N.J. plant production is 14 in. x 35-5/16 in., shingles per square 68.
Application procedure: For starter use full shingle with 6" trimmed off the butt edge. Overhang the eave ½" and secure with 4 nails 1½" up from bottom edge.
Apply shingles starting at the bottom of the roof, working across and up. This will blend shingles from one bundle into the next and minimize any normal shade variation.
Apply first course with full shingle even with starter course. Offset second course 4½" and third course 7½". Continue up the roof using the sequence-full shingle, 4½" offset, 7½" offset or any combination to achieve a random effect.

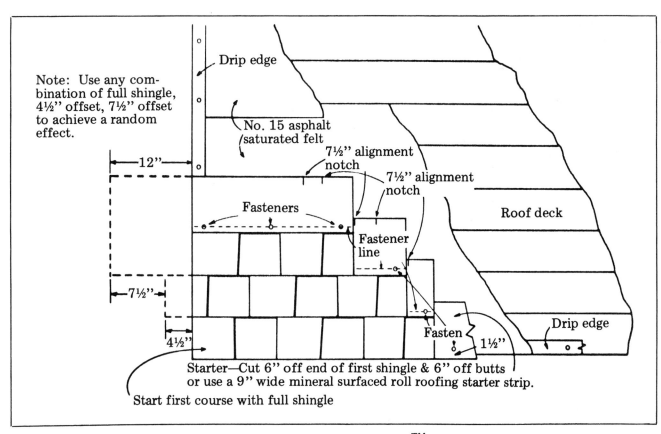

Johns-Manville - The Woodlands™ shingle
Starting procedure

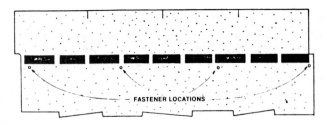

### RAMPART™ (Fiber Glass Shingle)
Shingle specifications: 20-year guarantee, U/L type class A and wind resistant.
Approximate weight: 225 lbs. per square.
Dimensions: Length 36¼ in.; width 12 in. (See note on Manville, N.J. plant).
Exposure: 5 in.; Headlap 2 in.
Bundles per square: 3
Shingles per square: 80
Nails per shingle: 4
1¼ in. nails per square. 2 lbs. 1½ in. nails per square: 2¼ lbs.
Hip and Ridge Shingles: 9 in. x 12 in., 240 shingles (2 bundles) per 100 lineal feet with 5 in. exposure. (West coast only). Other areas can cut the Rampart shingle into fourths for ridge. Manville, N.J. plant production is 12¼" x 36¼" with 5-1/8" exposure.
Clean White Algae Resistant shingle is available in southern areas.
Application procedure: For starter use full shingle with 5" trimmed off the butt. Overhang eave ½" and secure with four nails, 1½" up from bottom edge.
Apply shingles starting at the bottom of the roof, working across and up. This will blend shingles from one bundle into the next and minimize any normal shade variation.

Start each succeeding course after the first, up to and including the fourth with a shingle from which an additional 9" has been removed. Start the fifth course with a full shingle and repeat pattern. For the proper exposure, use the alignment offset to align each course with the top of the course below.

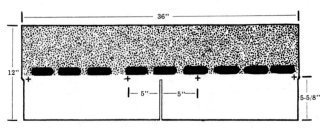

### FIRE-KING® (Fiber Glass Shingle)
Shingle specifications: U/L type class A and wind resistant
Approximate weight: 260 lbs. per square
Dimensions: Length 36 in.; width 12 in. (See note on Manville, N.J. plant).
Exposure: 5 in.; Headlap 2 in.
Bundles per square: 3
Shingles per square: 80
Nails per shingle: 4
1¼ in. nails per square: 2 lbs.; 1½ in. nails per square: 2¼ lbs.
Hip and Ridge shingles: 9" x 12", 240 shingles (2½ bundles) per 100 lineal feet with 5 inch exposure.
Manville, N.J. plant production is 12¼" x 36" with 5-1/8" exposure.
78 shingles per square.
Bright white fungus resistant surface available in southern areas.

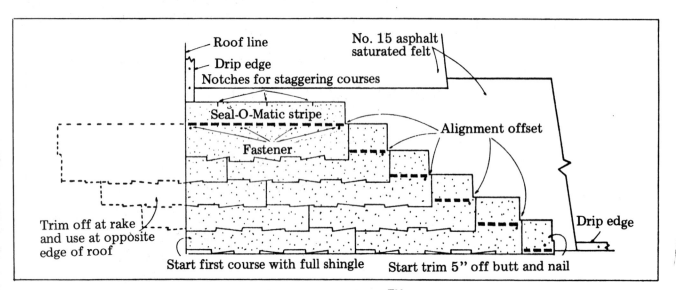

Johns-Manville - Rampart™ shingle
Starting procedure

Application procedure: Standard two tab methods: Same as the Highlands® shingle.

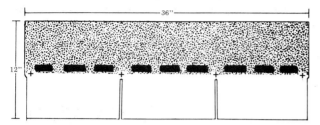

FIBER GLASS/3-TAB

Shingle specifications: U/L type class A and wind resistant.
Approximate weight: 225 lbs. per square.
Dimensions: Length 36 in.; width 12 in. (See note on Manville, N.J. plant).
Exposure: 5 in.; Headlap 2 in.
Bundles per square: 3
Shingles per square: 80
Nails per shingle: 4
1¼ in. nails per square: 2 lbs.; 1½ in. nails per square: 2¼ lbs.
Hip and Ridge shingles: Cut from the same Fiber Glass shingles. 3 - 12'' x 12'' tabs.
Manville, N.J. plant production is 12¼'' x 36'' with 5-1/8 in. exposure and 78 shingles per square.
Application procedure: Standard 3-tab methods.

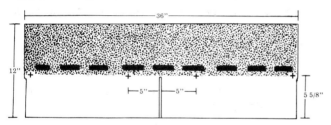

THE GARRISON®
(Fiber Glass Organic Shingle)

Shingle specifications: U/L type class A and wind resistant.
Approximate weight: 325 lbs. per square.
Dimensions: Length 36 in.; width 12¼'' in.
Exposure: 5 in.; Headlap 2 in.
Bundles per square: 4
Shingles per square: 80
Nails per shingle: 4
1¼ in. nails per square: 2 lbs., 1½ in. nails per square: 2¼ lbs.
Hip and Ridge shingles: 9 in x 12 in., 240 shingles (5 bundles) per 100 lineal feet with 5 in. exposure.
Application procedure: Standard two tab methods: Same as the Highlands® shingle.

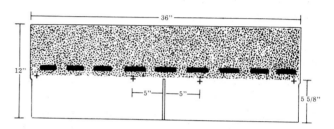

THE HIGHLANDS®

Shingle specifications: U/L type class C and wind resistant.
Approximate weight: 300 lbs. per square.
Dimensions: Length 36 in.; width 12¼ in.
Exposure: 5 in.; Headlap 2 in.
Bundles per square: 4
Shingles per square: 80
Nails per shingle: 4
1¼ in. nails per square: 2 lbs. 1½ in. nails per square: 2¼ lbs.
Hip and Ridge Shingles: 9 in. x 12 in. 240 shingles (3 bundles) per 100 lineal feet with 5 inch exposure.

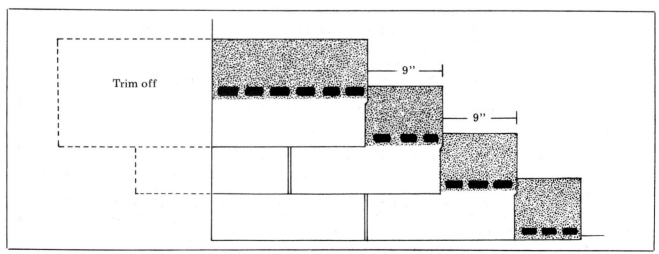

Johns-Manville - The Highlands® shingles
Starting procedure

Manville, N.J. plant production is 12¼" x 36" with 5-1/8" exposure and 78 shingles per square.

Application procedure: For starter use full shingle with tabs trimmed off butt. Overhang eave ½" and secure with four nails, 1½" up from the bottom edge.

Apply shingles starting at the bottom edge of the roof, working across and up. This will blend shingles from one bundle into the next and minimize any normal shade variation.

Start each succeeding course after the first, up to and including the fourth with a shingle from which an additional 9" has been removed. Start the fifth course with a full shingle and repeat pattern.

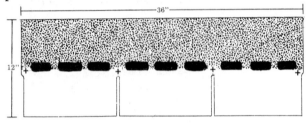

STANDARD SEAL-O-MATIC®

Shingle specifications: U/L type class C and wind resistant.

Approximate weight: 235 lbs. per square.

Dimensions: Length 36 in.; width 12 in. (See note on Manville, N.J. plant).

Exposure: 5 in.; Headlap 2 in.

Bundles per square: 3

Shingles per square: 80

Nails per shingle: 4

1¼ in. nails per square: 2 lbs.; 1½ in. nails per square: 2¼ lbs.

Hip and Ridge shingles: 9 in. x 12 in. 240 shingles (2 bundles) per 100 lineal feet with 5 in. exposure.

Manville, N.J. plant production is 12¼" x 36" with 5-1/8" in. exposure and 78 shingles per square.

Application procedure: Standard 3-tab methods.

# Owens-Corning Roofing Materials

Owens-Corning Fiberglas Corporation
Home Building Products Division
Fiberglas Tower
Toledo, Ohio 43659

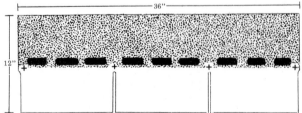

THE GLASTAR™ 20$_S^S$ SHINGLE
20-year guarantee
THE GLASTAR™ 25$_S^S$ SHINGLE
25-year guarantee

Shingle specifications for both: U/L type class A and wind resistant

Approximate weight: Glastar 20$_S^S$ - 205 lbs. per square. Glastar 25$_S^S$ - 260 lbs. per square

Dimensions: Length 36 in.; width 12 in.

Exposure: 5 in.; Headlap 2 in.

Bundles per square: 3

Shingles per square: 80

Nails per shingle: 4

1¼ in. nails per square: 2 lbs.; 1½ in. nails per square: 2¼ lbs.

Hip and Ridge: Cut from the same Glastar shingles.

Application procedure: Standard 3-tab methods.

*Reference Section*

# Tamko Roofing Materials

Tamko Asphalt Products, Inc.
General Offices: 601 North High Street
Joplin, Missouri 64801

Plants:
601 N. High Street, P. O. Box 1404, Joplin, Missouri 64801
U. S. Highway 183, N. city limits, P. O. Box 326, Phillipsburg, Kansas 67661
Kaul Industrial Park, 2300 35th Street, P. O. Box 2149, Tuscaloosa, Alabama 35401
Additional warehouses:
3009 East 17th Street, P. O. Box 4422, Kansas City, Missouri 64127
15 North Boston Avenue, Tulsa, Oklahoma 74103
5300 East 43rd Avenue, Denver, Colorado 80216
6101 North 16th Street, Omaha, Nebraska 68110

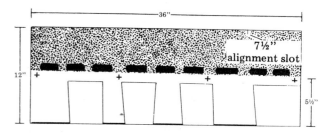
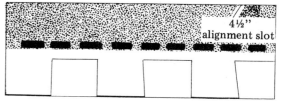

HERITAGE SHINGLE
Shingle specifications: 25-year warranty, U/L type class C and wind resistant

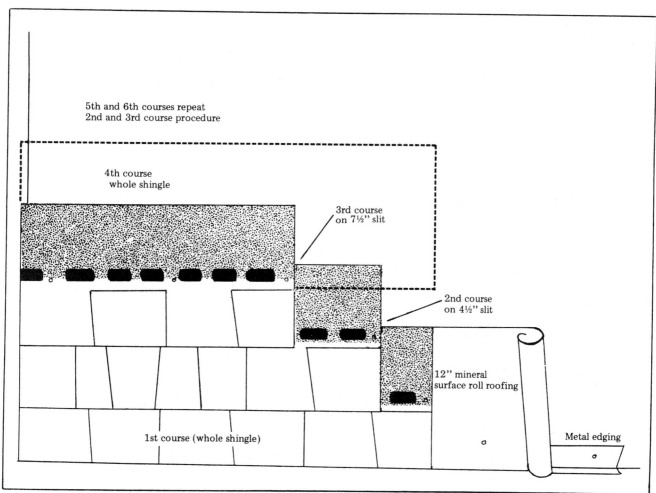

Tamko Heritage shingle
Starting procedure

185

Approximate weight: 328 lbs. per square
Dimensions: Length 36 in.; width 12 in.
Exposure: 5 in.; Headlap 2 in.
Bundles per square: 4
Shingles per square: 80
Nails per shingle: 4
1¼ in. nails per square: 2 lbs.; 1½ in. nails per square: 2¼ lbs.
Hip and Ridge shingles: Use hip and ridge (standard) of the same color.
Application procedure: Use 12 in. wide mineral surface roll roofing for starter course and nail securely. Cut the first piece of starter about 7½ feet long so that the joints will not coincide with the joints in the first course of shingles. Any additional starter pieces should be cut in 6 ft. or 9 ft. lengths.

First shingle will be a whole shingle aligned with the starter. Second shingle will be laid with the bottom edge even with the top of the saw tooth of the underlying shingle to obtain the 5 in. exposure. To offset it correctly the right end should be set at the first alignment slit from the right end of the underlying shingle (4½ in. or 7½ in.). The third shingle applies the same way, with the right end on the first slit from the right. If the second shingle was set on the 4½ in. slit, the third shingle should be set on the 7½ in. slit or vice versa. This is as far as the step-in goes, as the fourth shingle will start over again with a whole shingle. The fifth and sixth shingles will be a repeat of the second and third shingles but they will not need to be in the same order. If the second shingle was set on the 4½ in. slit, the fifth can be set on the 7½ in. slit.

Strike a chalk line about every six courses to keep the shingles straight across the roof. Shingles can be laid from either the right side or the left side of the roof. When laying shingles from the left hand side (as in the drawing) use the alignment slits farthest from the rake edge. When laying shingles from the right side use the alignment slits *closest* to the right edge.

Hip roof application: Strike a vertical chalk line from the hip and ridge junction at the top to the eave to use as a guide for starting the 1st, 4th, 7th, etc. courses.

Shingle specifications: 15-year warranty, U/L type class C and wind resistant
Approximate weight: 237 lbs. per square
Dimensions: Length 36 in.; width 12 in.
Exposure: 5 in.; Headlap 2 in.
Bundles per square: 3
Shingles per square: 80
Nails per shingle: 4
1¼ in. nails per square: 2 lbs.; 1½ in. nails per square: 2¼ lbs.
Hip and Ridge: Use standard hip and ridge shingles.
Application procedure: Application method is similar to the standard 3-tab shingle. Lay at random pattern but be sure that the end joints are offset as per manufacturers recommendation on the bundle wrapper.

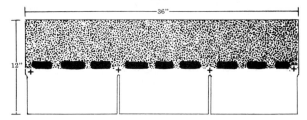

**240 SEAL DOWN SHINGLE**

Shingle specifications: 15-year warranty, U/L type class C and wind resistant
Approximate weight: 234 lbs. per square
Dimensions: Length 36 in.; width 12 in.
Exposure: 5 in.; Headlap 2 in.
Bundles per square: 3
Shingles per square: 80
Nails per shingle: 4
1¼ in. nails per square: 2 lbs.; 1½ in. nails per square: 2¼ lbs.
Hip and Ridge shingles: use standard hip and ridge shingles or cut from the 240 Seal Down shingle.
Application procedure: Standard 3-tab methods.

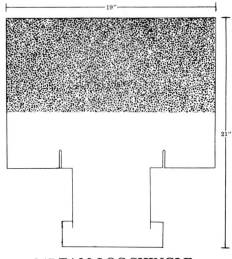

**245 TAM-LOC SHINGLE**

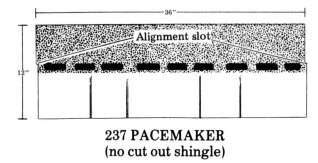

**237 PACEMAKER**
(no cut out shingle)

*Reference Section*

Shingle specifications: 17-year warranty, U/L type class C
Approximate weight: 245 lbs. per square
Dimensions: Length 19 in.; width 21½ in.
Bundles per square: 3
Shingles per square: 108
Nails per shingle: 2
1¼ in. nails per square: 1½ lbs.; 1½ in. nails per square: 1¾ lbs.
Hip and Ridge shingles: use standard hip and ridge shingles or cut from the 240 Seal Down shingle.

147 Hip and Ridge shingles: 9" x 12" 240 shingles (2 bundles = 1 square) per 100 lineal feet with 5 in. exposure.

44 Mineral surface valley strip: Length 36 ft., width 18 in. Comes in one 36 in. roll (two 18 in. rolls).

22 Mineral surface starter strip: Length 36 ft., width 9 in. Comes in one 36 in. roll (four 9 in. rolls).

# Index

**A**

Additions
  Cutting in a valley . . . . . . . . . 76
  Repairing leaks. . . . 102-106, 108
  Wall flashing. . . . . . . . . . . . . . 93
Advertising . . . . . . . . . . . . 144-149
  Door knob hangers . . . . . . . 148
  Giveaway items . . . . . . . . 147
  Letters . . . . . . . . . . . . . . . . 147
  Newspaper advertising . . . . . 145
  Newspaper publicity . . . . . . 146
  Planning your advertising . . . 145
  Printed matter . . . . . . . . . . 149
  Signs . . . . . . . . . . . . . . . . . 146
Attic vents, reroofing around 90, 91
Architect 70, shingle, Bird . . . . 160

**B**

Backing down
  T-lock tie-in . . . . . . . 48-51, 54-56
  3-tabs, 5 inch pattern. . . . . 57, 58
Beginning a roof . . . . . . . . . . . . . 13
Bird roofing . . . . . . . . . . . . 160-162
Blind valleys . . . . . . . . . . . . . 71, 72
Butting-up . . . . . . . . . . . . . . 27-32
Butt-nailing . . . . . . . . . . . . . . . 133

**C**

Celotex roofing . . . . . . . . . 162-167
Certain-teed roofing . . . . . 167-171
Chalklining hips for ridging . . . . 77
Chimneys . . . . . . . . . . . . . . . 62-64
  Flashing and sealing . . . . . . . 63
  Repairing leaks . . . . . . . . . . . 64
  Sealing . . . . . . . . . . . . . . . . . 63
Cold application . . . . . . . . . 35, 101
Concealed nailing . . . . . . . . . . . 95
Condensation, metal . . . . . . . . 118
Continuous wall flashing . . 129, 130
Contract, proposal and . . . . 155-156
Coverage (see Exposures)
Crew management . . . . . . . . . . 141
Cricket, fireplace . . . . . 61, 62, 122
Cutting hip ridge from 3-tabs 78, 79
Cutting in a valley on an addition 76

**D**

Dimensional shake
  shingle, Celotex . . . . . . . . . . . 162

Double coursed application . . . . 133
Dutch weave pattern . . . . . 124, 125

**E**

Equipment, roofing . . . . . . . . 8-12
Estimating
  Butt-up reroof . . . . . . . . . . . . . 28
  Estimating & measuring 134-141
  Nail amounts . . . . . . 113, 160-187
  Replacing bad sheathing . . . . 35
  Ridge . . . . . . . . 115, 137, 160-187
  Shakes . . . . . . . . . . . 114, 115, 127
  Wall exposures, shake. . 127, 133
  Wood . . . . . . . . . . . . 114, 115, 118
Exposures
  Asphalt . . . . . . . 13, 20, 21, 23, 28
  Shakes . . . . . . . . . . . 113, 127, 133
  Wood . . . . . . . . . . . . 113, 114, 118

**F**

Felt underlay. . 34, 86, 125-128, 159
  Asphalt shingle roof . 34, 86, 159
  Shake shingle roof . . . . . 125-128
  Low slope requirements . . . . 159
Fire-chex shingle, Celotex . . . . 164
Fire Guard shingle, GAF . . . . . 178
Fire-king shingle,
  Johns-Manville. . . . . . . . . . . . 182
Fireplace and chimney
  flashing . . . . . . . . . . . . . . 60-64
  Building a cricket . . . . . . . . . 62
  Flu caps . . . . . . . . . . . . . . . . 64
  Making watertight corners . . 61
  Metal saddle flashings . . . . . . 60
  Repairing leaks . . . . . . . . 63, 64
  Step flashing . . . . . . . . . . . . 61
  Wooden cricket
    flashing . . . . . . . . . 61, 62, 122
Fish mouthed ridge . . . . . . . . . . 82
Five inch pattern . . . . . . . . . . . . 21
Flashing,
  wall 34, 92, 93, 120, 121, 129, 130
Flashing, metal
  Tying into flat roofs . . . . . . . 34
Flashing,
  vent 35, 86-91, 121, 122, 126, 131
  Sealing . . . . . . . . . . . . . . . . 89

Flintkote roofing . . . . . . . . 172-173
Flu cap . . . . . . . . . . . . . . . . . . 64
Four inch pattern . . . . . . . . . . 24
Freeze back . . . . . . . . . . . 6, 86
  Felt underlay . . . . . . . . . . . 159
FRS shingles, Celotex . . . . . . 165
Fry roofing . . . . . . . . . . 174-177
Full lace valley . . . . . . . . . 65, 66

**G**

GAF roofing . . . . . . . . . . 177-180
Gauging
  3-tabs . . . . . . . . . . 20, 21, 23
  Wood . . . . . . . . . . . . . 114-116
Glassguard shingle,
  Certain-teed . . . . . . . . . . . 169
Glasstex shingle, Certain-teed 169
Glastar shingles,
  Owens-Corning. . . . . . . . . . 184
Garrison shingle,
  Johns-Manville . . . . . . . . . 183
Gravel roofs
  Tear off and reroof . . . . . 33-35

**H**

Hail damage, repairing . . . . . . 94
Half lace valley . . . . . . . . . 66, 67
Half pattern (see Six inch pattern)
Hallmark Shangle, Certain-teed 167
Hatchet, roofers . . . . . 10, 11, 112
Headlap . . . . . . . . 13, 27-32, 41
  3-tab shingles . . . . . . 13, 27-32
  Roll roofing . . . . . . . . . . . . 28
  T-lock shingles . . . . . . . 28, 41
Heritage shingle, Tamko . . . . . 185
Highlands shingle,
  Johns-Manville. . . . . . . . . . 183
Hip pads
  How to make . . . . . . . . . . . 11
  Where to order . . . . . . . . . 112
Hook blades . . . . . . . . . . . . . . 10
Hours per square chart . . . . . . 139

**I**

Independence shangle,
  Certain-teed . . . . . . . . . . . 168

**J**

Jet, Bird Wind seal shingle . . . . 162
Johns-Manville roofing . . . 181-184
Jointed inside corner . . . . . . . 133

**K**

Knife, hook blade . . . . . . . . . . . 10

**L**

Laced valleys . . . . . . . . . . . 65-67
Laying out the shingles
  Three tabs . . . . . . . . . . . . . 83
  T-lock . . . . . . . . . . . . . . . . . 84
  Shakes . . . . . . . . . . . . 128, 129
  Wood . . . . . . . . . . . . . . . . 120
Leaks. . . . . . . . . . . . . . 6, 7, 63,
  64, 69-74,81,88-90, 102-111, 130
  Fireplace . . . . . . . . . . 63, 64, 110
  Flat roof . . . . . . . . . . 106, 108
  Middle of slope. . . . 107, 110, 111
  Ridge . . . . . . . . . . . . . . . . . 81
  Sidewalls . . . . . . . . . . . . . 130
  Valleys . . . . 69, 74, 102-106, 111
  Vents . . . . . . 88, 90, 103, 107-110
Loader, shingle. . . . . . . . . . . 8, 9
Lock shingle, Fry . . . . . . . . . 177
Low pitch roofs
  Felt application (low slope) . . 159
  Roll roofing . . . . . . . . . 99, 101
Low slope application. . . . . . . 159

**M**

Mansard, nail requirements . . . 159
Mark 25 shingle, Bird . . . . . . . 160
Measuring . . . . . . . . . . . 136-138
  Gable roof . . . . . . . . . 137, 138
  Hip roof . . . . . . . . . . . 136, 137
  Triangular roof . . . . . . . . . . 138
Metal edging . . . . . . . . . . . 96-98
Metal ridge, reroofing over . . . 82
Metal valley . . . . . . . . . 66-68, 73-75
  Smooth metal . . . . . 66-68, 73
  W shaped formed
    valley . . . . . . . 67, 68, 73-75
Metal wall flashing . . . . . 119, 121
Mitered outside corner . . . . . . 133

## N

Nail bag .................................. 10
Nail remover and cutter ................ 95
Nail stripper ............................. 9
Nailing instructions
  How to hold big headed nails .38
  Mansard or steep roofs ..... 157
  Metal edging .............. 96, 97
  Ridge (asphalt) .............. 80
  Shakes ............. 113, 114, 133
  Shingle replacing ............ 95
  Three tab shingles
    (new work) ....... 19, 21, 24
  Three tab shingles
    (reroofs) ............. 24, 30
  T-lock shingles ............ 37, 38
  Valleys .................... 69, 72
  Wood shingles ......... 113, 114
Nails, pound per square
  Asphalt shingles ........ 160-187
  Wood shingles and shakes ..113
NCO shingle, Flintkote ......... 173
Ninety pound valley .......... 66, 72
Nor'easter shingle, GAF ........ 180

## O

Operating a roofing
  company .................. 134-143
  Crew & equipment
    management ............. 141
  Estimating ............. 134-141
  Hours per square chart ..... 139
  Measuring .............. 136-138
  Profit ................... 140, 141
  Reroofing guideline chart ... 135
  Starting a new company 142, 143
  Training programs ..... 141, 142
Overlap .......... 27-32, 41, 100
  Roll roofing ............ 28, 100
  Three tab ............... 27-32
  T-locks ................ 28, 41
Owens-Corning roofing ....... 184

## P

Pacemaker shingle, Tamko .... 186
Pad, hip ..................... 11, 112
  How to make one ........ 11
  Where to order one ...... 112
Patterns (3-tabs)
  Five-inch pattern .......... 21
  Four-inch pattern ......... 24
  Six-inch pattern ......... 18, 23
  Pattern selection .......... 18
Pitch, roof ................. 15, 33
  Determining the exact pitch ..33
Preparing roof for ridge..... 80, 81
Profit ..................... 140, 141
Proposal and contract ..... 155-156

## R

Rake
  Trimming the shingles ....... 41
  Precutting the shingles ..... 40
Rampart shingle,
  Johns-Manville ........... 181
Replacing shingles .......... 94-96
  Asbestos ................. 95
  Shakes ................... 95
  Slate .................... 95
  Three tabs ............... 94
  Tile ..................... 95
  T-locks ............... 95, 96
  Wood .................... 95
Reroofing chart ............. 135
Reroofing over old metal ridge ..82
Reroofs ....... 14, 18, 27-35, 41, 42
  Butting up .............. 27-32
  3-tab shingles ...... 14, 18, 27-35
  T-lock shingles .......... 41, 42
Ribbon courses .............. 85
  Removal ................. 85
Ridge ... 77, 82, 115, 122, 124, 137
  Asphalt shingles ......... 77-82
  Chalklining for alignment ...77
  Cutting hip ridge from
    3-tabs ................. 78, 79

Estimating amount
  needed ............... 115, 137
  Fishmouthed ridge ......... 82
  Nailing ................ 77, 80
  Nails .................... 77
  Preparing roof for ridge... 80, 81
  Reroofing over old metal ridge 82
  Ridge leaks ............... 81
  Shakes ............... 131, 132
  Wood roofs .......... 122, 124
Roll roofing ....... 34, 35, 100, 101
Roof failures
  Shakes ............... 129, 130
  Three tabs .......... 27, 28, 31
  T-locks .................. 84
  Wood ............... 113, 114
Roof jacks, metal ......... 11, 12
Roof pitch ............... 15, 33
Roof seat ............. 119-120
Roofing equipment .......... 8
Roofing, general information ...6
Roofing hatchet ...... 10, 11, 112
Roofing history ............. 5
Roofing terms ............ 16, 17
Roofing tools .............. 8-12
Roofing, your future in ...... 5-6
Rustic Crestwood shingle, Fry .175
Rustic shingle, Celotex ...... 164

## S

Saddle flashing ............. 60
Safe-Lox shingle, Celotex ..... 166
Saf-T-Lok shingle, Certain-teed 171
Salesmanship ........... 150-154
Scaffold, roof ..... 11, 12, 119-121
Sealing a vent .............. 89
Selling your services ..... 144-156
  Advertising .......... 144-149
  Proposal and contract ...155-156
  Salesmanship ........ 150-154
Serrated pattern ...... 85, 86, 116
  Laying a serrated pattern ...116
Shadow course .............. 85
  Removal ................. 85
Shake look (wood) ...... 124, 125
Shakes ............. 93, 125-133
  Estimating .... 114, 115, 134-141
  Exposures ............ 113, 127
  Felt underlay ...... 125-128, 133
  Flashing, wall (step) 93, 129, 130
  Nail selection ............. 113
  Red Cedar Shake chart ..... 127
  Ridge ............. 125, 131, 132
  Shingle replacing ......... 95
  Steep roofs ...... 128, 129, 133
  Valleys .............. 128, 129
  Vents ............ 126, 130, 131
Sheathing, spaced ....... 114, 115
Shingle repair .............. 94
Shingle replacing .......... 94-96
  Asbestos ................. 95
  Shakes ................... 95
  Slate .................... 95
  Three tab ................ 94
  Tile ..................... 95
  T-locks ............... 95, 96
  Wood .................... 95
Shingle selection ..... 84, 85, 113
Shingle types ............ 13, 14
Shingles, stacking on roof ..... 14
Shurlok shingle, Flintkote .... 173
Sidewall exposures
  Shakes ............... 127, 133
  Wood ................... 118
Sierra shingle, Flintkote ...... 172
Silicone sealer .............. 64
Single coursed application ... 133
Six inch pattern ......... 18, 23
Smooth valley ............ 66-72
Sovereign shingle, GAF ...... 179
Spot repairs ................ 94
Square, definition of ........ 17
Squaring up ............ 32, 33
Staggered pattern (wood) ..85, 86
  Laying a staggered pattern ..116
Stairstepping ............ 18-21
Staple gun ............. 117, 118
Starter course

3-tab shingles ......... 19, 27, 29
T-lock shingles .......... 36, 39
Roll starter ............... 39
Starter-finish shakes ....... 126
Starting a roofing company 142, 143
Steep roofs
  Getting more traction ....... 38
  Wood roofs ........... 119-121
  Scaffold ............... 11-12
  Shake roofs .......... 129, 133
Step flashing
  Fireplace ................ 61
  Shakes .......... 93, 129, 130
  Three tab ................ 92
  T-locks ............... 92, 93
  Wood ........... 93, 120, 121
Storm King shingle, Celotex ...165
Straightening out shingles
  Three tab shingles .... 21, 22, 25
  T-lock shingles ........ 41, 42
Straight-up method ... 20, 23, 24, 31
Strip Seal shingle, Certain-teed 169
Stripper, nail ..... 9, 112, 116, 117
  Where to order .......... 117
Suburban Thatch shingle, Fry .174

## T

Tamko roofing materials ..185-187
Tam-Loc shingle, Tamko ..... 186
Tear-offs ................ 99, 100
  Wood shingles ............ 99
  Composition .......... 99, 100
Terms used in roofing ..... 16, 17
Thatch look (wood) ..... 124, 125
Three tab shingles ....... 18-35
  Determining roof pitch ..... 33
  Five inch pattern .......... 21
  Four inch pattern ....... 24, 25
  Gravel tear off and reroof ..33-35
  Horizontal alignment ....... 24
  How to apply square lines ...33
  Laying out shingles ....... 83
  Nailing instruction & pattern .19
  Reroofing (butting-up) ... 27-32
  Selecting the pattern ...... 18
  Shingling straight up (butt-up)32
Six inch pattern
  stairstep method ........ 18-20
Six inch pattern, straight up
  method ................ 23
Straightening a five inch
  pattern ................ 22
Straightening horizontal lines .26
Straightening vertical lines 25, 26
Trimming rakes .......... 98
Tie-ins ................. 43-59
  Difficult T-lock ........ 54-56
  Three tab five inch pattern,
    new roof ............ 56-59
  Three tab six inch pattern,
    new roofs ............ 43-46
  Three tab six inch pattern,
    reroofs ............... 47
  Shingling around a large
    dormer ............. 53, 54
Timberline shingle, GAF..... 177
Tips, roofing ........... 83-101
  Felt underlay ............ 86
  Laying out shingles ..... 83, 84
  Metal edging .......... 96-98
  Roll roofing .......... 100, 101
  Shingle repair ........... 94
  Shingle replacement ..... 94-96
  Shingle selection ....... 84-86
  Step flashing .......... 90-94
  Tear-offs ............ 99-100
  Trimming rakes ......... 98
  Vents ................ 86-91
Tite-On shingle, GAF ...... 179
T-locks .................. 36-42
  Dimensions of different
    brands ............. 164-184
  Improper application ..... 15
  Nailing instructions ...... 37
  Offsetting overlap ...... 41, 42
  Precutting rake shingles .... 40
  Replacing shingles ..... 95, 96
  Shingling around vents ...86, 87

3-tab shingles ......... 19, 27, 29
T-lock shingles .......... 36, 39
Roll starter ............... 39
Starter-finish shakes ...... 126
Starter course .......... 36, 39
Step flashing ........... 92, 93
Straightening out ....... 41, 42
Trimming rakes .......... 98
Toe board ..... 11, 12, 39, 119-121
  When to use ............. 39
  Wood roofs .......... 120-121
Tools, roofing ...... 8-12, 95, 112
  Hatchet ............. 10, 11, 112
  Hip pad ............. 11, 112
  Nail remover ............ 95
  Nail stripper ......... 9, 112
Tool belt .................. 10
Traditional shake shingle,
  Celotex ................ 163
Training programs ..... 141, 142
Trimming rakes ............ 98

## U

Underlay, felt
  Asphalt shingles ....... 34, 86
  Cutting 18 inch felt ...... 133
  Low slope application .... 157
  Shake shingles ..... 125-128

## V

Valleys ................. 65-76
  Blind valleys .......... 71, 72
  Cutting in a new valley
    (additions) ............ 76
  Full lace (woven) ...... 65, 66
  Half lace .............. 66, 67
  Leak prevention ....... 69-71
  Leaks ..... 69-74, 102-106, 111
  Metal condensation ...... 118
  Replacing a valley ..... 74-76
  Selecting the right valley ..72-74
  Shakes .............. 128, 129
  Smooth valleys (90 lb. or
    metal) .............. 66-72
  Wood roofs .......... 118, 119
  W-shaped formed
    valley ....... 67, 68, 73-75
Variation of shingles .20, 24, 27, 32
Vent flashings
  Leaks ........... 103, 107-109
  Reroofs ............ 35, 88-91
  Sealing ................ 89
  Shake roofs ...... 126, 130, 131
  Wood roofs ......... 121, 122
Ventilation ............... 114

## W

Wall exposures
  Shakes .............. 127, 133
  Wood .................. 118
Wall flashing
  Flat roofs .............. 34
  Shakes .......... 93, 129, 130
  Three tabs .............. 92
  T-locks ............... 92, 93
  Wood ........... 93, 120, 121
Wooden cricket .......... 61, 62
Wood shingles .......... 112-125
  Application ............ 115
  Estimating114, 115, 118, 134-141
  Exposures ........ 113, 114, 118
  Dutch weave pattern ..124, 125
  Leaks ........... 104, 106, 107
  Nail selection .......... 113
  Red Cedar shingle chart ..118
  Reroofs ............. 113, 114
  Ridge .............. 122-124
  Staple gun ......... 117-118
  Steep roofs ......... 119-121
  Shingle replacing ........ 95
  Shingle selection ........ 113
  Tearing off ............ 99
  Tools .................. 112
  Valleys ............. 118, 119
  Vent flashings ....... 121, 122
  Ventilation ............ 114
  Wall flashing ........ 93, 121
Woodlands Roof shingle,
  Johns-Manville .......... 180
Woodman shingle, Bird ..... 161
Woodtex shingle, Certain-teed.169
Woven corners ........... 133
Woven valley ........... 65, 66

# Other Practical References

**National Construction Estimator**
Current building costs in dollars and cents for residential, commercial and industrial construction. Prices for every commonly used building material, and the proper labor cost associated with installation of the material. Everything figured out to give you the "in place" cost in seconds. Many time-saving rules of thumb, waste and coverage factors and estimating tables are included. **512 pages, 8½ x 11, $16.00. Revised annually.**

**Estimating Tables for Home Building**
Produce accurate estimates in minutes for nearly any home or multi-family dwelling. This handy manual has the tables you need to find the quantity of materials and labor for most residential construction. Includes overhead and profit, how to develop unit costs for labor and materials and how to be sure you've considered every cost in the job. **336 pages, 8½ x 11, $21.50**

**Building Cost Manual**
Square foot costs for residential, commercial, industrial, and farm buildings. In a few minutes you work up a reliable budget estimate based on the actual materials and design features, area, shape, wall height, number of floors and support requirements. Most important, you include all the important variables that can make any building unique from a cost standpoint. **240 pages, 8½ x 11, $12.00. Revised annually**

**Rafter Length Manual**
Complete rafter length tables and the "how to" of roof framing. Shows how to use the tables to find the actual length of common, hip, valley and jack rafters. Shows how to measure, mark, cut and erect the rafters, find the drop of the hip, shorten jack rafters, mark the ridge and much more. Has the tables, explanations and illustrations every professional roof framer needs. **369 pages, 8½ x 5½, $12.25**

**Stair Builders Handbook**
If you know the floor to floor rise, this handbook will give you everything else: the number and dimension of treads and risers, the total run, the correct well hole opening, the angle of incline, the quantity of materials and settings for your framing square for over 3,500 code approved rise and run combinations—several for every 1/8 inch interval from a 3 foot to a 12 foot floor to floor rise. **416 pages, 8½ x 5½, $12.75**

**Wood-Frame House Construction**
From the layout of the outer walls, excavation and formwork, to finish carpentry, and painting, every step of construction is covered in detail with clear illustrations and explanations. Everything the builder needs to know about framing, roofing, siding, insulation and vapor barrier, interior finishing, floor coverings, and stairs. . . complete step by step "how to" information on what goes into building a frame house. **240 pages, 8½ x 11, $11.25. Revised edition**

**Rough Carpentry**
All rough carpentry is covered in detail: sills, girders, columns, joists, sheathing, ceiling, roof and wall framing, roof trusses, dormers, bay windows, furring and grounds, stairs and insulation. Many of the 24 chapters explain practical code approved methods for saving lumber and time without sacrificing quality. Chapters on columns, headers, rafters, joists and girders show how to use simple engineering principles to select the right lumber dimension for whatever species and grade you are using. **288 pages, 8½ x 11, $14.50**

**Contractor's Guide to the Building Code**
Explains in plain English exactly what the Uniform Building Code requires and shows how to design and construct residential and light commercial buildings that will pass inspection the first time. Suggests how to work with the inspector to minimize construction costs, what common building short cuts are likely to be cited, and where exceptions are granted. **312 pages, 5½ x 8½, $16.25**

**Contractor's Year-Round Tax Guide**
How to set up and run your construction business to minimize taxes: corporate tax strategy and how to use it to your advantage, and what you should be aware of in contracts with others. Covers tax shelters for builders, write-offs and investments that will reduce your taxes, accounting methods that are best for contractors, and what the I.R.S. allows and what it often questions. **192 pages, 8½ x 11, $16.50**

**Builder's Office Manual**
This manual will show every builder with from 3 to 25 employees the best ways to: organize the office space needed, establish an accurate record-keeping system, create procedures and forms that streamline work, control costs, hire and retain a productive staff, minimize overhead, and much more. **208 pages, 8½ x 11, $13.25**

**Manual of Professional Remodeling**
This is the practical manual of professional remodeling written by an experienced and successful remodeling contractor. Shows how to evaluate a job and avoid 30-minute jobs that take all day, what to fix and what to leave alone, and what to watch for in dealing with subcontractors. Includes chapters on calculating space requirements, repairing structural defects, remodeling kitchens, baths, walls and ceilings, doors and windows, floors, roofs, installing fireplaces and chimneys (including built-ins), skylights, and exterior siding. Includes blank forms, checklists, sample contracts, and proposals you can copy and use. **400 pages, 8½ x 11, $18.75**

**Reducing Home Building Costs**
Explains where significant cost savings are possible and shows how to take advantage of these opportunities. Six chapters show how to reduce foundation, floor, exterior wall, roof, interior and finishing costs. Three chapters show effective ways to avoid problems usually associated with bad weather at the jobsite. Explains how to increase labor productivity. **224 pages, 8½ x 11, $10.25**

**Basic Plumbing with Illustrations**
The journeyman's and apprentice's guide to installing plumbing, piping and fixtures in residential and light commercial buildings: how to select the right materials, lay out the job and do professional quality plumbing work. Explains the use of essential tools and materials, how to make repairs, maintain plumbing systems, install fixtures and add to existing systems. **320 pages, 8½ x 11, $17.50**